Relationship with the Land

Hugh Hammond Bennett, Aldo Leopold, and the Future of the Conservation Land Ethic

Edited by Mark Anderson-Wilk

Soil and Water Conservation Society
Ankeny, Iowa

Soil and Water Conservation Society
945 SW Ankeny Road, Ankeny, IA 50023
www.swcs.org

Printed in the United States of America.
5 4 3 2 1
ISBN 978-0-9769432-6-6

Library of Congress Cataloging-in-Publication Data

Relationship with the land : Hugh Hammond Bennett, Aldo Leopold, and the future of the conservation land ethic / edited by Mark Anderson-Wilk.
  p. cm.
 ISBN 978-0-9769432-6-6 (pbk.)
 1. Conservationists--United States--Biography. 2. Bennett, Hugh H. (Hugh Hammond), 1881-1960. 3. Leopold, Aldo, 1886-1948. 4. Naturalists--United States--Biography. 5. Conservation of natural resources. 6. Land use. 7. Ecosystem management. 8. Soil conservation--United States. I. Anderson-Wilk, Mark.
 S415.R 395 2009
 333.95'160922--dc22

                    2008054159

The Soil and Water Conservation Society is a nonprofit scientific and professional organization that fosters the science and art of natural resource management to achieve sustainability. The society's members promote and practice an ethic that recognizes the interdependence of people and their environment.

Photos on back cover. Top: This cow is eating green *Sericea lespedeza* from Dr. H.H. Bennett's hand, Spartansburg, South Carolina. Photo by B.D. Robinson, courtesy of the USDA Natural Resources Conservation Service. Bottom: Aldo Leopold at his desk, 424 University Farm Place, Madison, Wisconsin, late 1930s, from *Aldo Leopold: The Man and His Legacy*, ed. Thomas Tanner, published by the Soil and Water Conservation Society, 1987.

Table of Contents

Part 4. The Future of the Conservation Land Ethic

Preface

Hugh Hammond Bennett, the first chief of the USDA Soil Conservation Service, and Aldo Leopold, best known as the author of *A Sand County Almanac*, are "larger than life" figures who continue to cast their influence not only on conservationists working in the United States but on all sorts of people around the world concerned with the sustainability of soil, water, and other precious natural resources.

Conservation, it can be argued, is achieved by the dedication and actions of countless ordinary individuals. Natural resource conservation is a quality that defines the relationship between individual land managers and their land. One would think, then, that conservation would not naturally be a "celebrity" profession, but—in relative terms—Bennett and Leopold have achieved iconic status at least within the conservation movement.

Though this book focuses on the ethical ideas of Bennett, Leopold, and others, some description is also provided of their personalities, which were important factors in helping communicate their ideas. Maurice G. Cook, for example, describes the role Bennett's charisma and ability to network played in the successes he achieved. Melville H. Cohee calls Leopold a leader whose "mold was lost when he was born."

Part 1 of the book highlights Hugh Hammond Bennett's preeminent role in the US soil conservation movement. As the first chief of the Soil Erosion Service/Soil Conservation Service (now called the Natural Resources Conservation Service) and founder of the Soil Conservation Society of America (now called the Soil and Water Conservation Society), Bennett was clearly the chief architect of the movement's institutional structures. But his ethical commitment to the land and his unique approach to caring for it led him to also be a greater-than-life hero of the land and a magnetic visionary who continues to touch every new generation of soil conservationists.

Part 2 focuses on Aldo Leopold and the development of ideas that constitute his conservation land ethic. Chapters are included by leading Leopold scholars—Susan Flader and Curt Meine. In the final piece of part 2, Mel Cohee, writing in 1987 on the occasion of the 100th anniversary of Leopold's birth, shares his experience working with Aldo Leopold and Hugh Hammond Bennett in the early days of the US soil conservation movement.

Part 3 provides examples of the development of conservation thought related to land stewardship and natural resource values. Clarence J. Glacken and Carl C. Taylor summarize the historical background of conservation philosophy and ethics. Pieces by Paul B. Sears and Firman E. Bear demonstrate the US conservation movement at mid-20th century: Sears expresses his dismay over the depletion of natural resources in making war munitions; Bear beams a broad, patriotic spotlight on the many forms of conservation that contribute to making America beautiful and abundant. Paul A. Herbert, William Voigt Jr., and James W. Giltmier express their views that include a concern about the role of government in private land conservation.

Finally, R. Neil Sampson, Fred B. Samson and Fritz L. Knopf, and Tony Prato offer their perspectives on the developing concept of ecosystem management. An interesting addition to this discussion is found in Bryan G. Norton's reading of Leopold as an early developer of the adaptive ecosystem management concept (see part 4).

Also in part 4, Norm Berg and Pete Nowak question the utility of the land ethic in affecting real change. Another chapter looks at some of the past philosophical conflicts and potential future opportunities in the conservation movement.

Part 4 includes two invited pieces written specifically for this volume. Norva Y.S. Lo considers the land ethic from an international perspective. The land ethic is in the process of a "makeover;" it has recently been redubbed by some as the "earth ethic," reflecting the interconnected global reality of the world's resources and human activity. Lo provides some context for a global land ethic related to ecosystem health and the malaise of "affluenza." Kristin Shrader-Frechette, in perhaps the chapter in the volume most challenging to the conventional soil conservationist perspective, suggests that the concept of private property needs to be reconsidered for the land ethic to be realized.

The volume concludes with a chapter by Wendell Berry that argues "people cannot be adequately motivated to care for land by general principles or by incentives.... They must feel that the land belongs to them, that they belong to it." The title of the book, *Relationship with the Land*, connects with both Berry's concept of living with the land and Leopold's extension of community to include the land.

The following themes can been found running through this collection of pieces representing over 60 years of conservation thought and diversity of perspectives:

- *Ethics and Experience.* Individual actions and personal stories can have wide, transformative effects for people and the land.
- *Ethics and Responsibility.* Global, ecosystem-based stewardship requires resources to be treated as something beyond property rights—more like part of the living community of which Aldo Leopold and Wendell Berry write.
- *Ethics and Knowledge.* The land ethic and what we've learned about ecosystem services together should guide us in land management policies and practices.

- *Ethics and Application.* The land ethic is most meaningful when closely tied to pursuing what works—that is, what has the greatest practical ability to result in real, measurable conservation.

Ethics and Experience

Wheeler McMillen reminds us that, though Hugh Hammond Bennett was a master of developing soil conservation institutions, he believed that ultimately soil conservation is a "one-man job"—that is, soil conservation must be carried out by each individual farmer or farm owner. On a similar note, Leopold's insisted that the farmer must be the one to do conservation "because he is the only person who resides on the land and has complete authority over it" (see chapter by Curt Meine).

Bennett and Leopold are both excellent illustrative examples of personal stories that result in widescale positive impacts.

Phoebe Harrison recounts a young Hugh Hammond Bennett's first personal experience with soil erosion and conservation technique. The experience naturally connects with Bennett's later professional career; one man, driven by an ethic for the land, kicked off a large-scale movement to defeat the "evil" of soil erosion.

Susan Flader provides a biography of Aldo Leopold that provides insight into the connection between his personal experience and the development of his career and philosophical ideas. Flader describes Leopold's evolving relationship with the land and ideas in as personal a context as within his letters to his mother and his wife. Later in his life, his personal experience as land steward at his "shack" and farm property in "Sand County," Wisconsin, deepened and personalized his land ethic beliefs.

Joseph Hart presents the story of the Haugen brothers. The Haugen family farm, in the Coon Creek, Wisconsin, watershed, was part of the nation's first conservation project in the 1930s. The Haugens have witnessed first hand decade after decade of changing trends in soil erosion, farming practices, and conservation approaches. Farmers such as the Haugens at once transform the land and are transformed by it.

R. Neil Sampson uses his own experience, from growing up on a farm through his career as a natural resource conservationist, to demonstrate the development and enrichment of personal land ethics. He concludes by urging readers "to participate to the fullest … applying those shared values to the daily tasks we face." Or, as Firman E. Bear puts it, "conservation starts with man."

Ethics and Responsibility

Leopold is known for advancing the idea that the land is part of the community, and thus, the soil, water, plants, and animals that make up the land deserve to be treated as such; that is, humans have an ethical obligation to the biotic whole. As Susan Flader explains, Leopold argued "the case for a sense of obligation to the community going beyond economic self-interest." Or, as James W. Giltmier puts it, "sustainable stewardship is good because it ignores short-term profits for long-term, continuing benefits."

Firman E. Bear, writing in 1960, bemoans that "if a man wants to plow up a hill—that is his license, as far as present laws are concerned.... If a man wishes to cover the best agricultural land in the nation with pavements ... let his own poor judgment guide him. We are still lacking in legal code and even in politic conscience, as regards an obligation to posterity in this respect."

R. Neil Sampson, writing in 1992, equates the soil conservation movement to the civil rights movement: "The United States has established a national policy that says we should prevent soil abuse that robs our future generations, but then has been willing only to implement that policy up to a point... there's a long way to go, just like there was in 1960 with civil rights, before we really address the attitudes and values people hold."

Shrader-Frechette takes up the case of the relationship between private property/resource ownership and political, social, and economic power and opportunity. Based on her analysis, she concludes that "limiting property rights in natural resources might provide a foundation for avoiding the land destruction."

Counterpoints and critiques are also provided. Writing in 1965, William Voigt Jr., with some disapproval, uses the example of the Soil Conservation Service to show how public agencies have grown in their scope as their roles and responsibilities as public guardians of the land have expanded. James W. Giltmier, writing in 1990, responds to some of the questions environmentalists have posed as to the ability of traditional land managers to adequately ensure the sustainability of the land.

Ethics and Knowledge

The Soil and Water Conservation Society (SWCS) is known for the science of soil and water conservation. The SWCS membership by and large represents the technical and program side of private land conservation—the scientists, the technology developers and implementers, the technical advisors, the educators, the policy makers, the program administrators. So why a book on the philosophy and ethics of conservation? A close reading of the role of the society in the conservation movement since its early days in the 1940s shows that SWCS has consistently embraced both the hard and soft sides of conservation, with recognition that conservation cannot advance without advances in both the science and the art of conservation.

This book addresses science's role in conservation as related to environmental ethics.

Writing in 1950, Paul B. Sears points out that "we have been applying science most powerfully to ... ends that are certainly not conservative." He provides the example of "the elaboration of consumers' goods, with attendant marketing pressure and without much reference to raw materials."

This analysis is further elaborated in chapters by Y.S. Lo and Wendell Berry. The consumer culture that has only expanded since 1950 has provided more emphasis on how to fully utilize resources and maximize productivity to meet the world's growing consumption demands than on how to minimize the growing consumption needs and ensure sustainability.

The discussions of ecosystem management in this book (see Sampson, Samson and Knopf, and Prato) attempt to improve the connection between science and practice by advancing models to apply scientific understanding to environmental concerns in a way that better matches the growing global, ecosystem-based land ethic.

The recent downturn in the world economy and the growing nonrenewable energy and global warming crises hold some promise of "forcing our hand" toward reduced consumption and increased conservation, but the realities of changes in policy and practice are yet to be seen.

Ethics and Application

Norm Berg's contribution, written in 1998, brings his policy experience to bear on the land ethic and finds, even after decades of land policy and conservation program development, a meaningful application of land and water stewardship still wanting.

The land ethic by itself has been identified as inadequate as a motivator for practicing conservation methods.

Ted L. Napier, recipient of the 2008 Hugh Hammond Bennett Award, recently noted that "most farmers already have internalized positive attitudes toward stewardship of land and water resources." Based on decades of socioeconomic research and work with farmers, Napier concludes the issue is that "farmers approach soil and water conservation at the farm level from a business perspective and cannot be expected to adopt and use a farm production system that will not produce profits."

In his piece in this collection, Pete Nowak agrees. He notes that farmers by and large are already familiar with the land ethic, but that they also have "something that may be called a market ethic, a family ethic, a community ethic, and so on." The typical farmer then is not opposed to conservation but is compelled to make decisions as a business person and someone responsible for a family. It is not reasonable to expect the farmer to voluntarily uphold environmental protection above his/her other interests.

Napier goes further to point out that "there will simply be not enough money available from the US government to subsidize grain farmers to the level required to motivate them to operate their farm businesses in a manner that would be 'environmentally friendly;'" thus, some form of regulation is required (Napier's article can be found in the January/February 2009 issue of the *Journal of Soil and Water Conservation*.)

The answer to the question of what policies would result in the greatest environmental benefits at a feasible cost is complex and still far from fully realized.

An interesting observation that can be made in rereading these selections from the conservation library is that even the older pieces hold up well with time. This is a testament to conservation principles, which—though contextual and a product of cultural perspective—have lasting interest and usefulness. While the programs and techniques of conservation have matured over time as the science of conservation has improved (think of the over-reliance on terraces that concerned Hugh Hammond Bennett, or the tile that the Soil Conservation Service helped install to drain farms of natural wetlands, or the burning of wheat stubble and the straightening of streams R. Neil Sampson discusses in his chapter, only later to be shown as counterproductive and reversed as recommended practices), conservation *ideas* have not so much as improved over time but become more complex and contextual as they are applied to specific programs and practices.

What is striking about this collection is that the authors are, for the most part, not formally educated as philosophers or ethicists. They represent a range of occupations, which is an indication of the land ethic's perceived relevance across a wide swath of fields and peoples.

Hugh Hammond Bennett and Aldo Leopold may have both been larger than life, but the principles they stood for and spent their careers working for have implications not only for land managers and conservationists but for the globe's entire population, which depends on the health and services of resource-rich ecosystems for continued sustenance and survival.

Mark Anderson-Wilk

Part 1
Hugh Hammond Bennett and
the Soil Conservation Movement

Weaver of Principles into a Science of Conservation for the Land

Phoebe Harrison

Editor's note. This piece, written shortly after Hugh Hammond Bennett's death on July 7, 1960, notes Bennett's role in the formation of the Soil Conservation Service and the Soil Conservation Society of America (now called the Natural Resources Conservation Service and the Soil and Water Conservation Society, respectively).

Nearly seventy years ago the young boy Hugh Hammond Bennett stood in the mud of a North Carolina field and watched his father build a small bank across the slope.

"You do that to hold the soil," he remarked.

"Yes, son. It helps some to stop the gullies."

The boy looked around, up and down the hilly, worn out farm.

"Gullies, yes, there they are. We can make more banks."

When his life ended, the man Hugh Hammond Bennett had to his credit the stopping of uncounted millions of gullies, design and fabric for a clean and cared for rural landscape in his own beloved country and fanning out over the world, and a shining pattern for leadership and vision regarding use and conservation of our only true heritage, our soils.

We are now reminded of the slow early years of his struggle for recognition of the land problem as he saw it so vividly. For long there was doubt as to whether the people would ever catch the real picture of their rapidly declining agricultural base. We see him, a young man, stalking the hills and valleys for soil samples, erosion causes, production figures, people suffering for need of good ground in which to grow their food. For the true and convincing story of soil ruin and waste he searched the land itself, while out of his own conviction grew his remarkable capacity for instilling his faith and conviction in others.

Once the way was cleared and he was permitted to speak out, the driving force of his campaigning for soil conservation was so compelling as to be unique in agricultural history. When in his prime and as we first knew him well, it early became apparent that he would succeed in awakening Americans to the dangers of continuing waste of the land resource. In assembly halls across the nation, country people and town people alike, alert to his every word, testified to the success of his grassroots awakening. This man loved the land of America. He had an affinity for the soil, the waters, and all that grew therefrom. He had the rare quality we can only

3

call perception for things and trends not seen by most, and he truly loved America. And he was so human his listeners forgot that he was a noted scientist.

Later, after he had performed that near heroic feat of creating a new science—soil and water conservation—and had organized a new kind of government agency to cope with waste of soil, then did we see the practical side of the man. He had a dream but he was no dreamer. In a matter of weeks, he brought men of many sciences together to build the Soil Conservation Service. He sent his technicians into the field to start the biggest face-lifting job in history. He was called the one-idea scientist, but his idea had many facets. Classify and use the land according to its capabilities, plan the farms for complete conservation, acre by acre, to give farmers something to work by, manage the water and the soil as one, never relax vigilance in using the soil the conservation way—these are but a few of the principles he wove into a science of conservation for the land.

With amazing foresight he looked always into the future while he spread lasting conservation use and treatments over the land. Alert always to anything that might threaten the permanence and soundness of soil conservation for his country, he met it with hard facts and feet planted firmly on the ground.

Hugh Hammond Bennett was the founder of the Soil Conservation Society of America. It was part of his dream, along with an ever productive and beautiful landscape. We take great pride in our founder, in the honors heaped upon him while he lived, in his ineradicable imprint on our lives and our land. We shall miss him from our ranks.

ORIGINAL PUBLICATION DETAILS

By Phoebe Harrison, USDA Soil Conservation Service.
Harrison, Phoebe. 1960. Hugh Hammond Bennett. *Journal of Soil and Water Conservation* 15(5):198.

Ground Lost and Gained in 75 years of Conservation at Coon Creek

Joseph Hart

Editor's note. This piece was written on the occasion of the 75th anniversary celebration of the first federal conservation demonstration project at the Coon Creek watershed in Wisconsin, April 25, 2008. Hugh Hammond Bennett and Aldo Leopold, along with other conservation leaders of the day, played a role in the landmark project at Coon Creek.

One spring morning, the back room at the local Eagles club in the tiny southwestern Wisconsin town of Coon Valley was filled to capacity with farmers, outdoorsmen, scientists, conservationists, and politicians. Over coffee and donuts, this unusual alliance gathered to pay homage to a humble element that ties together all their varied interests: soil.

Interrupted by occasional camera flashes, local and national dignitaries, including senators, congressmen, and the chief of the USDA Natural Resources Conservation Service, Arlen Lancaster, stood at the podium extolling the virtues of the soil conservation methods pioneered in the region. "Every farmer in the nation has a tie to the Coon Creek watershed," Lancaster said. "You changed the course of our history."

Seventy-five years ago—during one of the worst environmental and economic disasters this nation has ever experienced—modern soil conservation was born in the surrounding farmland of Coon Creek. The watershed was a living laboratory to test soil science, conservation techniques, and private-public cooperation. Then, as now, collaboration among scientists and landowners resulted in dramatic gains: In a matter of years, a denuded landscape devastated by erosion was healed. And in the process, modern conservation was invented.

Fighting the "National Menace"

Farmers who lived through the "dirty '30s" faced double trouble: The stock market crash and the depression that followed dispossessed thousands who lost their farms through bank foreclosure. Those who held onto their homes watched their topsoil disappear, literally blown away in the dust storms of the 1930s.

Starting with the drought of 1931, windstorms whipped up dust from parched fields into clouds of soil that blotted out the sun for days at a time. The storms were choking. Farmers wore face masks and smeared their nostrils with Vaseline in an attempt to breathe outdoors, and entire farmsteads were buried in dirt.

By 1934, the US government estimated that 1.4×10^7 ha (3.5×10^7 ac) of cultivated croplands had been "essentially destroyed" by soil erosion, while 4.0×10^7 ha (1.0×10^8 ac) had lost "all or most of the topsoil" (USDA 1934).

This environmental crisis came as no surprise to Hugh Hammond Bennett, who had been proselytizing for decades about the "national menace" of soil erosion.

Bennett writes that his epiphany on the subject came all the way back in 1905, when he was working as a soil surveyor for the USDA. He and his partner were making a survey of a tract of Virginia landscape when they observed side-by-side fields demonstrating the dramatic results of what is now known as sheet erosion.

"The slope of both areas was the same," Bennett later wrote. "The underlying rock was the same. There was indisputable evidence that the two pieces had been identical in soil makeup. But the soil of one piece was mellow, loamy, and moist enough even in dry weather to dig into with our bare hands. We noticed this area was wooded, well covered with forest litter, and had never been cultivated. The other area, right beside it, was clay, hard and almost like rock in dry weather. It had been cropped a long time" (Cook n.d.).

The lessons of these field observations were as profound as they were simple: The time-honored farming techniques of Europe, imported by immigrants to the United States, were destroying the nation's soils—and by extension, its ability to feed itself.

The next 30 years only proved the point, and when the Dust Bowl hit, with its devastating loss of topsoil, Bennett was poised to do something about it. He approached President Franklin D. Roosevelt with a proposal. "He asked for funding to do demonstration projects to see how best to address the erosion problems," explains Wisconsin State Conservationist Pat Leavenworth. "Roosevelt was skeptical and said he would base additional funding on that initial project. And it had to be in a limited time period. So whether the Coon Creek watershed succeeded or failed was the basis for our future conservation programs."

In 1933, faced with this challenge, a budget of $5 million and the blessing of the president, Bennett launched a demonstration project designed to prove the lessons he learned back in 1905—that "vegetation, with minimal engineering, could check the runaway erosion that was ruining America's breadbasket" (Anderson 2002).

According to Renae Anderson, public affairs specialist with the USDA Natural Resources Conservation Service in Madison, Wisconsin, Bennett settled on the Coon Creek watershed in part because it lay near to the south of La Crosse, Wisconsin, home to a federal erosion experiment station, and in part because farmers in the watershed were open to trying something new. In other regions, anti-government sentiments prevailed, but in Coon Creek, farmers were either open enough—or desperate enough—to accept government help and payments for soil conservation.

"Bennett designated it 'Project No. 1,' and it became the first watershed project in the nation. It was 22 miles long and nine miles wide, encompassing 92,000 acres straddling three counties, with outlet directly to the Mississippi River" (Anderson 2002).

It didn't hurt that Coon Creek farms were undeniably scarred by erosion. The watershed is part of the Driftless region, an unglaciated landscape of springs and sinkholes, cold-water trout streams, and extremely steep wooded slopes that stretches from western Wisconsin into eastern Minnesota and northeastern Iowa.

Following old-world farm methods, European settlers had stripped the countryside of trees and planted neat-as-a-pin rows of oats, wheat, and other crops. After mere decades of such treatment, fields were gashed by gullies too deep to cross with a wagon. Topsoil washed downhill into the bottomlands where it settled 3 to 5 m (10 to 15 ft) deep in some places, turning cold-water freshets into sluggish sloughs.

In short, the land was in crisis. Bennett assembled a team that included regional director Raymond H. Davis, conservationists Herbert A. Flueck and Marvin F. Schweers, forester/wildlife management specialist Aldo Leopold, biologist Ernest G. Holt, economist Melville H. Cohee, and agricultural engineer Gerald E. Ryerson.

It was an immersive, collaborative experience. "Aldo Leopold writes about sitting around the campfires at night talking about the day—what worked and what didn't," Leavenworth says. "They were all living together onsite." Together, they devised a system of soil conservation that relied on terracing, contour planting, pasturing, crop rotations, and replanting: all of which have since become the hallmarks of modern conservation.

No less important were the techniques used to implement these reforms. Farmers then as now were suspicious of self-proclaimed "experts." Bennett and his team could offer farmers the incentive of government payments, but that in itself wouldn't be enough to convince wary farmers to convert. Instead, the scientists had to demonstrate that erosion could be halted, and that yields could be improved. They had to show that their notions would actually work.

"That's a real key to working with private landowners, versus public lands," Leavenworth says. "You have to get them to believe in you."

A Healing Landscape

Ernest Haugen remembers the day the work crews with their bulldozers began constructing the first terraces in the Coon Creek conservation project on his father's farm. Today, the 87-year-old still farms with his brother Joseph on their father's spread, although in recent years they've reduced their stock to a half-dozen cows, and a neighbor cuts the hay on their terraced fields.

When Bennett's team began work on his father's farm, supported by a crew of unemployed men trucked in by the Civilian Conservation Corps, the success of the demonstration project was far from assured. Some of Haugen's neighbors signed up along with him, but many others held back. Even true believers figured maybe they'd go back to their traditional farming practices after the five-year government payout program ended.

Instead, farmers rapidly enlisted in the program as they witnessed its success. "After the first year, they had to put additions on their barns for the increased hay yields," says Leavenworth. Eventually, most of the watershed adopted some form of conservation, while Bennett and his followers, their case proven, spread the practices honed in Wisconsin across the country, and indeed, the world.

Today, the fruits of the project are evident, as Haugen can point out. Here and there, an old gully dating from the early part of the 20th century has grown over with grass and trees. Slopes that once were grazed have grown over with soil-protecting forests. Haugen points out a massive culvert at the bottom of one steep field. Years ago, he says, road crews installed the culvert to handle runoff from the field, but in his tenure on the farm, it has remained dry.

Stanley W. Trimble, a professor of geography at UCLA, has studied the Coon Creek watershed since the 1970s (e.g., Trimble 1975, 1981, 1983, 1999), and his research confirms what Haugen has observed in his lifetime: Soil resilience is improved, sediment is reduced, and water quality has improved. Every year, Trimble measures stream sediment against baselines set by researchers immediately following the completion of the federal project. By these investigations, he can report that the watershed is healing. A more visible measure, he says, is the return of the cold-water-loving native brook trout to the watershed. "Stream quality is so much better than it was; it's gone from brown trout to brook trout that can reproduce themselves."

In 2007, the project faced its greatest threat yet—a massive flood with rains so heavy that many in the area have come to call "the 1,000-year flood." Many of the erosion control structures built in the 1930s were designed to withstand a heavy rain of 10 to 15 cm (4 to 6 in) in a 24-hour period, says Sam Skemp, head of Wisconsin's Vernon County Natural Resources Conservation Service field office. After 38 cm (15 in) of rain fell, "I imagined all our conservation work was over." Instead, what he saw amazed him. "Fifty years ago, 15 inches of rain would have wiped out the town of Chaseburg," he says, referring to a town that lies on Coon Creek. "Instead, there were places it was hard to tell there had been a rain storm."

A Fragile Future

On a spring day, Ernest Haugen's farm is vibrant with life, glowing green in the electricity of threatening thunderstorms. In later years, the Haugen brothers added to the system of check dams and terraces, creating a whole-farm system of water control. The earth is springy, and acres of thriving alfalfa pastureland are divided by barbed wire strung along hand-hewed posts of black locust—some of which were planted by the Civilian Conservation Corps in the 1930s.

Walk over the crest of the hill, however, and you'll see a study in contrasts as striking as the one that set Bennett on the warpath back in 1905. The next field over is brown and lifeless, last year's corn traced in straight lines across the field. Haugen shakes his head. "They came in and got rid of all the terraces," he says with a frown.

These two fields illustrate what may turn into the nation's next great soil crisis. In the 1930s, most of the farmers in southwestern Wisconsin were converting from crops like oats and wheat to diversified, small-scale dairies—operations that were tailor-made for the kind of reforms that Bennett brought there. Farm plans called for longer crop rotations, contour strips of hay, and pasturelands, all crucial elements in a small dairy operation.

In the past few decades, however, family-owned dairies have been replaced by massive feedlots with tens of thousands of cows. Many farmers, like Haugen's neighbor, are converting conservation reserve land to cropland, a trend that is poised to accelerate as the government subsidies for biofuels drive up the price of corn.

Fortunately, conventional farm practices have improved since the 1930s. In particular, the advent of no-till farming has reduced soil erosion on croplands, albeit with an increased reliance on chemical herbicides. "It's not the disaster it could be if farmers were using conventional techniques," says Trimble.

If the loss of conservation land continues, however, it could pose a threat to the fragile balance regained since the 1930s.

References

Anderson, R. 2002. Coon Valley days. Wisconsin Academy Review 48(2).

Browning, G.M., M.H. Cohee, N.J. Fuqua, R.G. Hill, and H.W. Pritchard. 1984. Out of the dust Bowl: Five early conservationists reflect on the roots of the soil and water conservation movement in the United States. Journal of Soil and Water Conservation 39(1):6-17.

Cook, M.G. n.d. Hugh Hammond Bennett: The Father of Soil Conservation. Raleigh, NC: Department of Soil Science, North Carolina State University. http://www.soil.ncsu.edu/about/century/hugh.html.

Helms, D. 1992. Coon Valley, Wisconsin: A conservation success story. Readings in the History of the Soil Conservation Service. Washington, DC: USDA Natural Resources Conservation Service. http://www.nrcs.usda.gov/about/history/articles/CoonValley.html.

Leopold, A. 1935. Coon Valley: An adventure in cooperative conservation. American Forests 41(5):205-208.

Trimble, S.W. 1975. Response of Coon Creek watershed to soil conservation measures. In Landscapes of Wisconsin, ed. B. Zakrzewska-Borowiecki, 24-29. Washington DC: Association of American Geographers.

Trimble, S.W. 1981. Changes in sediment storage in the Coon Creek basin, Driftless Area, Wisconsin, 1853 to 1975. Science 214:181-183.

Trimble, S.W. 1983. A sediment budget for Coon Creek basin in the Driftless Area, Wisconsin, 1853–1977. American Journal of Science 283:454-474.

Trimble, S.W. 1999. Decreased rates of alluvial sediment storage in the Coon Creek basin, Wisconsin, 1975–1993. Science 285:1244-1246.

USDA. 1934. Yearbook of Agriculture. Washington, DC: USDA.

ORIGINAL PUBLICATION DETAILS

By Joseph Hart, Freelance Writer, Viroqua, Wisconsin.

Hart, Joseph. 2008. Ground lost and gained in 75 years of conservation at Coon Creek. *Journal of Soil and Water Conservation* 63(4):102A-106A.

Big Hugh's Message: One Man's Task!

Wheeler McMillen

The last time I saw Hugh Bennett, we were having lunch as guests at an agricultural experiment station in Brazil. There on separate missions, our meeting was accidental. We had little time for conversation, and before I finished my meal, I saw him, soil auger in hand, on his way to the field.

"Hugh," I had asked before he left, "after all your countless studies and observations, at home and abroad, what is the most important thing you have learned?"

Without hesitation he replied, "… that a man can make his soil better than what nature provided for his use."

In terms of constructive usefulness to mankind, Hugh Bennett was perhaps the greatest man of our century.

From the beginning of his concern with soil erosion, he saw one irreversible fact—only one man could effectively preserve soil fertility and that man had to be the one who owned or controlled the land. True, Bennett helped initiate legislative measures for federal assistance to fight erosion. Such help, he believed, should involve efforts to discover and disseminate facts that farmers could apply. True, he assembled a large staff of competent aides—soil scientists, engineers, foresters, and others. But he instructed his staff never to call groups of farmers into a room where instructions would be issued. He insisted that his men go, when invited, to individual farms and, after thorough observation, suggest what the owner could best do to avoid erosion and build up his soil.

Because no man could control rain and wind, or alter the tendency of water to run downhill, Bennett's workers devised soil-conserving practices—terraces, stripcropping, windbreaks, and other measures to thwart the power of the natural elements. Some federal contribution might help defray the costs, but the actual soil preservation was the job of the man responsible for the land.

Bennett's preachments still hold true. Erosion today is most apparent where farm owners, for whatever reasons, have chosen to ignore fundamental conservation procedures. The land is permanent, though its fertile topsoil may not be. Land ownership is not permanent. It changes from one generation to the next, or as buying and selling alters possession.

Three attitudes about land ownership prevail. One looks at land ownership as stewardship, temporary control with the implication that the steward is obliged to treat the land so its future productivity is assured.

Another might misinterpret Thomas Jefferson's expression, "the land is for the living," and freely plow the slopes that ought to be in grass and lose no sleep if a valuable part of the farm goes down the river, never again to be fruitful.

A third attitude among farmers, and no doubt the most prevalent one, falls somewhere between these two extremes. A farmer in this group realizes, given the time and money, that he ought to preserve his soil and improve his land. He might plant trees along the banks of his creeks and along fencelines without reducing his tillable acreage much. He might rotate his crops more judiciously and make better use of his livestock wastes and other organic materials.

Land rental, of course, can present problems for erosion control efforts. A renter naturally wants to reap maximum profits for his work. But the owner remains the man in control. He can write the rental contract so his land is protected from abuse. In so doing, his own interests and those of the renter can probably be served.

Men farm to provide livelihoods for their families, to educate their children, and to provide security in old age. Most farmers like their work. But for a century they have been the most underpaid, though most necessary, workers in the nation. The reason why is obvious. Too many acres have been cultivated, and the flood of foodstuffs has gone cheaply into the cities, where profits on transport, trade, processing, and distribution have yielded the vast capital upon which industries and cities have flourished.

Now is heard the outcry that topsoil losses endanger future food supplies and threaten civilization worldwide. Not many individual farmers are joining the doomsayers. They know their most suitable soils will continue to produce the crops appropriate to them. But most farmers, even though they may accept as fact the assertion that more than one-third of cropland fertility is being lost faster than it is being replaced, will continue their struggle to make a living in this generation. Their efforts to keep fertility from blowing away or washing down the creeks will aim much more at their hopes for next year than for the concerns of future generations.

People will talk and worry, but we must not expect an organized outcry from consumers wishing to be taxed a little more to preserve or renew the soil that feeds them. Though his efforts can be aided by research, information, and some guidance, the individual farmer will long be the one man responsible for keeping the soil productive. That was part of the wisdom in Hugh Bennett's original plans. The farm owner is the *one man* in charge.

I believe what Bennett told me in that last encounter: Man can make the soil more productive than it was when taken from nature. But that will result only from many one-man jobs.

ORIGINAL PUBLICATION DETAILS

By Wheeler McMillen, Editor, *Farm Journal* (retired), Lovettsville, Virginia.

McMillen, Wheeler. 1982. Big Hugh's message: One-man's task! *Journal of Soil and Water Conservation* 37(4):195.

The Legacy of Hugh Hammond Bennett

Maurice G. Cook

T his year [2001] marks the 120th anniversary of the birth of Hugh Hammond Bennett. It's appropriate to review his marvelous legacy and apply it to the issues of today. Why was Hugh Hammond Bennett successful? How did he galvanize the soil conservation movement? Are the elements of his success relevant to the soil and water conservation scene today?

Bennett had many personal qualities that helped him succeed. In 1905, Bennett had definite ideas about soil erosion but not ideas that everyone agreed with. When sent to Alaska to keep him quiet, one of his qualities being perseverance, rather than being deterred, Bennett worked all the harder collecting data as a soil scientist and recording his observations. Remember, he "discovered" the soil erosion process in 1905. It was at least 20 years before his ideas began to be accepted.

Bennett's other strong personal quality is that fact he was a "people person." He enjoyed most getting out on the land and talking with people. Although he traveled a lot, he didn't like hotels. His common practice was to drive up to a farmhouse, introduce himself, help with the evening chores, and then stay overnight with them.

Bennett and the Soil Conservation Movement

What enabled Hugh Bennett to galvanize the soil conservation movement? In addition to his admirable personal qualities and his keen skills, Bennett was politically astute. He was a master of the networking process. By the late 1920s, his wake-up call about erosion was beginning to receive some attention, due in large part to the publication, *Soil Erosion—A National Menace*. He gained disciples everywhere he went. One was A.B. Conner, director of the Texas Agricultural Experiment Station. He encouraged Mr. Conner to talk with an influential member of the Appropriations Committee in the US House of Representatives, James P. Buchanan, also of Texas. Bennett, of course, saw the opportunity to fund his efforts. After much conversation, Buchanan invited Conner to come to Washington and present the soil erosion issue to the entire committee. Conner replied, "You really need a person who knows the subject nationally and I know just the person—Hugh Bennett." Bennett appeared before the committee, and it wasn't long before $160,000 was allocated to establish 10 research stations in strategic locations around the country. They were devoted exclusively to soil erosion research. The only logical choice of a person to direct this research was Hugh Hammond Bennett.

This was an example of Bennett's awareness of the importance of timing. Several years later, in the early 1930s, he seized the opportunity provided by the Dust Bowl to advance his cause. It is hard to think of anything good about the Dust Bowl. But we probably can say that it hastened the adoption of a soil conservation program in the United States. It's doubtful that it would have happened as rapidly and as dramatically without Bennett's keen sense of timing.

Following networking and timing, there is the important third step of getting the legislation, the funding, and whatever is necessary to get the conservation task accomplished. Bennett had a knack for connecting with powerful and influential people. He related well with Harold Ickes, Secretary of the Interior, who was close to President Franklin Roosevelt. Bennett soon experienced some of the closeness of the President.

When Bennett and others conceived the idea of a local entity—namely, the soil conservation district—his closeness to power paid off. A national model for establishing conservation districts, called the Standard State Soil Conservation Districts Law, was developed with the intent that each state would adopt it. President Roosevelt wrote personally to all governors encouraging them to adopt this master plan. In his letter, the President made this memorable statement (probably written by Bennett): "The nation that destroys its soil destroys itself."

Would this progress have occurred without Bennett's political astuteness? We don't know, of course, but I doubt it would have occurred so universally and so effectively without it.

His high-level contacts were not limited to the United States. On his overseas assignments early in his career, he somehow managed to connect with people in high places. He was invited back to those places later in his career.

There was never any doubt that Bennett staunchly believed in what he said and did. He perceived soil erosion as an evil. And he knew how to correct that evil. I think it was his passion for the cause, undergirded by scientific data to support it, which won the confidence of so many people.

Charisma was probably not a necessary prerequisite for Bennett's success, but it certainly added to his effectiveness. Perhaps his best-known charismatic display was his presentation in 1935 before a US Senate Committee on Public Lands and Survey. It also reflects Bennett's skillful timing. On the day that he was testifying, Bennett knew that a dust storm was on its way toward Washington, DC. He cited data after data, all the while keeping one eye on the clock. It was like a drama unfolding there in the conference room. In the afternoon, the skies became dark. One senator remarked that a rainstorm must be brewing. Another commented that it looked like dust. Bennett replied: "You know, Senator, I think you are right. It does look like dust!" This sealed the success of Bennett's mission. Congress quickly passed Public Law 46, the first soil conservation act in the history of any nation.

Current Relevance of Bennett's Success

It can be argued, with some justification, that Bennett's success story is irrelevant today. True, he lived in a different time and different circumstances. The country is a lot different now than it was in the 1930s. Bennett had a predominant single issue, soil erosion, which is not the front-page conservation headline today. But there are many natural resource issues, and soil and water are at the heart of them.

We shouldn't forget about soil erosion. It is still an issue. The main human factor causing erosion is different. It is not plowing up and down hill with mules, but it is soil movement by bulldozers and heavy equipment. Erosion is not as much a down-on-the farm problem as it is a development problem, i.e., highway construction, building construction, and the like. I think Bennett would be in the middle of all that development, talking with homebuilders associations, transportation departments, and with others about prescribing erosion control and sediment control policies.

There are several critical water issues. Water supply and distribution is becoming more and more a significant concern. Water demands continue to grow, increasing the need for water allocation to various sectors. This creates competition for water. A classic current example of competition and controversy is occurring in the Klamath River Valley in Oregon. On the one hand, there is the suckerfish, an endangered species. On the other, there are farmers needing the water for irrigation. Which water use should have the higher priority? We are likely to see more of this kind of conflict.

Water quality, in both surface and ground water, is a major concern. There is the debate over agricultural versus nonagricultural impacts. Who should bear the cost of improving water quality, individual users or society, or both?

What would be Bennett's position on these serious water issues? I don't know, but I think he would be highly visible in addressing them. I also think he might tell us today to work hard at rebuilding a political commitment to conservation by using the best facts we can find and tell them in the most honest and compelling ways we know how. I also think he would urge us to speak in terms of ethics as well as economics. We must help people understand that when we inherited this land from our ancestors, we did not get some God-given right to steal it from our children.

Hugh Bennett was a special man for a special time in the history of the world, especially the United States. He was a man of integrity, simplicity, directness, and honor. He supplemented those qualities with the force, vigor, and intelligence of his own personality. He was many things: scientist, author, teacher, visionary, strategist, tactician, prophet—a great leader.

Louis Bromfield, well-known conservationist and contemporary of Bennett, said it well: "Hugh Bennett deserves the greatest honor from the American people as one of the greatest benefactors since the beginning of their history."

What a profound compliment! Bennett earned it. Long live the ideals and the spirit of Hugh Hammond Bennett!

ORIGINAL PUBLICATION DETAILS

By Maurice G. Cook, Professor Emeritus, North Carolina State University.

Cook, Maurice G. 2001. The legacy of Hugh Hammond Bennett. *Conservation Voices* 4(5):6-7.

Part 2

Aldo Leopold and the Land Ethic

Aldo Leopold and the Evolution of a Land Ethic

Susan Flader

Editor's note. This piece was written on the occasion of the Aldo Leopold centennial celebration held at Iowa State University in October 1986. That event resulted in the book *Aldo Leopold: The Man and His Legacy* (Soil and Water Conservation Society 1987), in which this chapter first appeared.

The last two decades have witnessed an explosion of interest in the ethical basis for people's relationship with their environment, as conservationists, scientists, political leaders, and philosophers have sought an undergirding for what some feared might otherwise be an ephemeral flurry of public concern about environmental quality. In the resulting literature on environmental philosophy, one name and one idea recur more frequently than any others: Aldo Leopold and his concept of a land ethic.

Leopold articulated his environmental philosophy most powerfully in "The Land Ethic," capstone of *A Sand County Almanac*, the slim volume of natural history essays for which he is best known today. But "The Land Ethic" and the shorter, lighter, more illustrative vignettes that help illuminate it draw their clarity, strength, and enduring value from a lifetime of observation, experience, and reflection.

At his death in 1948, Leopold was perhaps best known as a leader—*the* leader—of the profession of wildlife management in America. He was also a forester and is regarded as the father of the national forest wilderness system. In tracing the evolution of his land ethic philosophy, however, we must look beyond his interest in wildlife and wilderness to his lifelong effort to understand the functioning of land as a dynamic system, a community of which we are all members. Though his ethical philosophy was an outgrowth of his entire life experience, this quest for its origins focuses in particular on an aspect of his thought that has been less widely known, his concern about mountain watersheds and the problem of soil erosion.

Foundations

Aldo Leopold was born in 1887 in Burlington, Iowa, in his grandparents' home on a bluff overlooking the Mississippi River. "Lugins-land"—look to the land—the family called the place, reflecting their German heritage and love of nature. The patriarch was Charles Starker, trained in Germany as a landscape architect and engineer, who instilled his keen esthetic sense in his daughter Clara, Aldo's mother.

Clara had a pervasive influence on Aldo, eldest and admittedly the favorite of her four children, nurturing the esthetic sensitivity that would be so integral to his land ethic philosophy. Carl Leopold, Aldo's father, also had a formative influence on him. He was a pioneer in the ethics of sportsmanship at a time when the very notion of sportsman was taking shape in the United States.

Leopold frequently hunted with his father and brothers on weekends and arranged his classes so he would have weekday mornings free to cross the Mississippi for ducks or tramp the upland woods on the Iowa side. Summers he spent with the family at Les Cheneaux Islands in Lake Huron, fishing, hunting, and exploring. Despite the temptations of his avocation, he secured a remarkably good education in the Burlington schools, especially in English and history. Then, though he might have been expected to follow his father into management of the Rand & Leopold Desk Company, he went east to Lawrenceville Preparatory School in New Jersey, the Sheffield Scientific School at Yale, and eventually the Yale Forest School to prepare himself for a career in the new profession of forestry.

When he left for Lawrenceville in January 1904, it was with his mother's admonition to write frequently and "tell me everything." He wrote her from the train while passing through the mountains of Pennsylvania. He wrote again upon arriving in Lawrenceville, describing the lay of the land—"flat, but not so bad as I thought"—and yet again the next day, by which time he had taken an afternoon's tramp of some 15 miles and could pronounce himself "more than pleased with the country." That letter ran four Pages with detailed accounts of timber species, land use practices, and birds, as well as of the dormitory, his class schedule, problems in algebra and prospects in German, everything. From then on for a decade, through his school years in the East and his early career in the Southwest until after his marriage, Leopold would write at least weekly to his mother, less frequently to his father, brothers, and sister. By rough count he penned something on the order of 10,000 pages of letters. He had learned grammar and sentence structure in the Burlington schools, but he learned to write by writing. His mother insisted on it.

At Yale, Leopold's letters grew longer with more elaborate narrative and keen observation, the result of continued encouragement from home and more frequent and extended tramps through the countryside. He frequented special places—Juniper Hill, Marvelwood, Diogenes Delight, the Queer Valley—and wrote in detail of his encounters with foxes, deer, all manner of birds, moods, and weather. But when he began technical studies in forestry—dendrology, mensuration, silviculture, forest economics—the forays became less frequent as his dedication to the scientific challenges of the new profession absorbed his attention.

Leopold as Forester

Yale had established the first graduate school of forestry in the United States in 1900 with an endowment from the family of the nation's leading forester, Gifford Pinchot. The school promoted Pinchot's doctrine of scientific resource management and what Samuel Hays has characterized as the Progressive Era's "gospel of efficiency" (Hays 1959). When Leopold graduated with a master of forestry degree in 1909 and went to work for Pinchot in the U.S. Forest Service, he was one of an elite corps of scientifically trained professionals who would develop administrative policies and techniques for the fledgling agency charged since 1905 with responsibility for managing the national forests. He was assigned to the new Southwestern district embracing Arizona and New Mexico territories.

His private reflections, expressed in letters home and recalled in essays written some three decades later, indicate that he was thoroughly enamored with the stark beauty of the country. But his first publications, which soon began appearing in local Forest Service periodicals, reveal the extent to which he bought into the utilitarian emphasis of the forestry enterprise. The earliest was a bit of doggerel titled "The Busy Season" (Leopold 1911), which he inserted anonymously into a newsletter he edited for the Carson National Forest in New Mexico:

> There's many a crooked, rocky trail,
> That we'd all like straight and free,
> There's many a mile of forest aisle,
> Where a fire-sign ought to be.
>
> There's many a pine tree on the hills,
> In sooth, they are tall and straight,
> But what we want to know is this.
> What will they estimate?
>
> There's many a cow-brute on the range,
> And her life is wild and free,
> But can she look at you and say,
> She's paid the grazing fee?
>
> All this and more, it's up to us,
> And say, boys, can we do it?
> I have but just three words to say,
> And they are these: "Take to it."

As he was coming to grips with the myriad management problems of the Carson, Leopold was also struggling with a personal dilemma—how to win the attention of a Spanish senorita he had just met, when he was "1,000,000 miles from Santa Fe" and his rival was right on the scene. He called on his best resource, his skill at expressing himself in letters.

My dear Estella—
This night is so wonderful that it almost hurts. I wonder if you are seeing the myriads of little "Scharfchens volken" I told you about—do you remember the "little sheep-clouds"?—I have never seen them so perfect as they are tonight. I would like to be out in *our Canon*—I don't know how to spell it so you will

have to let me call it that—and see the wild clematis in the moonlight—wouldn't you? I wonder if we could find four more little bluebells for you to pin at your throat—they were beautiful that evening in the dim twilight as they changed with the darkness into a paler and more unearthly blue—and finally into that color which one does not see but *knows*—simply because they are Bluebells.

After six months of extraordinary letters, Estella decided on Aldo; a year later the two were married. For the rest of his life, Estella would inspire and respond to Leopold's special esthetic sensitivity, as his mother had earlier, keeping it alive even during the long years when he was otherwise preoccupied with more worldly, practical affairs. Six months after their marriage, Leopold was off settling a range dispute in a remote area when he got caught in a flood and blizzard. He became a victim of acute nephritis, a kidney ailment that nearly took his life. After 18 months of recuperation and long hours of reflection on how he might live whatever time remained to him, he was able to return to light office work at the district headquarters in Albuquerque.

It was at this juncture that Leopold became involved in wildlife conservation, developing a new emphasis on cooperative game management that became a model for Forest Service activity around the nation. Though it was a line of work of his own choosing, an outgrowth of his early avocation, he approached it not in an esthetic mode but in the scientific, utilitarian spirit of the Forest Service. He would promote game management as a science, modeled on the principles and techniques of forestry. Game could bring nearly as much income to the region as timber or grazing uses of the forests, he calculated, if enough effort, intelligence, and money were committed to develop the resource. Extermination of wolves, mountain lions, and other predatory species was a key element in his early program, one he would live to regret (Flader 1974).

By 1917 he had achieved national recognition for his successes in the Southwest and was beginning to publish his ideas on game management and forestry regularly in periodicals of nationwide circulation. From then until his death, he published frequently—never fewer than two articles a year, often more than a dozen. But for the next two decades his work was decidedly management-oriented. It was not until the last decade of his life that he began publishing the literary and philosophical essays for which he is best known today.

As one examines the contours of Leopold's life, it is ironic to find that the esthetic appreciation for wildlife that was so integral to his youth, so evident in his early letters, and so vital to his mature philosophic reflection was seemingly suppressed at mid-career, at least in his public persona, as he sought to develop a science and profession of wildlife management. The irony is compounded when one notes the extent to which he was pushing beyond traditional modes of thought in his understanding of the dynamics of Southwestern watersheds by the early 1920s, developing an interpretation of the functional interrelatedness of virtually all elements of the system save wildlife. It was as if his effort to achieve parity for game animals within the Forest Service model of professional management limited his ken at the same time that he felt less constrained about challenging orthodoxy on larger issues. Thus, it is to his thinking about watersheds and soil erosion that we must turn if we would understand the evolution of his concept of a land ethic—his capacity to think about the system as a whole.

Southwestern Watersheds and Moral Obligation

Ever observant, Leopold had noted the condition of watersheds on national forests in the Southwest from his earliest days as a forest assistant. His first assignment had been to map and cruise timber along the route of a proposed wagon road that would have to clamber high over the Blue Range on the Apache Forest in eastern Arizona because erosion had foreclosed all possibility of a more logical route through the once-lush bottomlands of the Blue River. When he was promoted in 1919 to assistant district forester and chief of operations in charge of business organization, personnel, finance, roads and trails, and fire control on 20 million acres of national forests in the Southwest, Leopold had an opportunity to observe conditions anew as he crisscrossed the forests on inspection trips. By then, erosion had washed out nearly 90 percent of the arable land along the Blue. Of 30 mountain valleys he tallied in Southwestern forests, 27 were already damaged or ruined.

Leopold thought long and hard about soil erosion. He wrote about the problem in his inspection reports, trying to make foresters in the field more cognizant of changes occurring before their eyes. He spoke about it to the New Mexico Association for Science in a strongly worded warning, "Erosion as a Menace to the Social and Economic Future of the Southwest," and reached for national attention to the problem through articles in the *Journal of Forestry*. He even grappled with the philosophical meanings of the phenomenon in a remarkable essay, "Some Fundamentals of Conservation in the Southwest," that was found in his desk after his death.

In "Some Fundamentals," written in 1923, Leopold probed the causes of the degradation that was reducing the carrying capacity of Southwestern ranges and considered the implications for human ethical behavior. "The very first thing to know about causes," he wrote, "is whether we are dealing with an 'act of God,' or merely with the consequences of unwise use by man" (Leopold 1979). Through analysis of evidence from tree rings, archaeology, and history, he concluded that the deterioration of organic resources in the Southwest could not be attributed simply to climatic change (an act of God). But the nature of the climate, characterized by periodic drought, had resulted in a delicately balanced equilibrium that was easily upset by man. Overgrazing, resulting from overstocking without regard to recurring drought, was the outstanding factor in upsetting the equilibrium, in his view. This conclusion that human beings bore responsibility for unwise land use led him to a philosophical discussion under the subtitle "Conservation as a Moral Issue."

He began with an admonition from Ezekiel:

Seemeth it a small thing unto you to have fed upon good pasture, but ye must tread down with your feet the residue of your pasture? And to have drunk of the clear waters, but ye must foul the residue with your feet?

Ezekiel seemed to scorn poor land use as something damaging "to the self-respect of the craft and society" of which one was a member. It was even possible, Leopold thought, that Ezekiel respected the soil "not only as a craftsman respects his material, but as a moral being respects a living thing." Leopold found support for his own intuitive feeling that there existed between man and earth a deeper relation than would follow from a mechanistic conception of nature in the organicism of the Russian philosopher P. D. Ouspensky, who regarded the whole earth and all its parts as possessed of soul or consciousness. "Possibly, in our intuitive perceptions, which may be truer than our science and less impeded by words than our philosophies," Leopold suggested, "we realize the indivisibility of the earth—its soil, mountains, rivers, forests, climate, plants, and animals, and respect it collectively not only as a useful servant but as a living being."

Realizing that this premise of a living earth might be too intangible for many people to accept as a guide to moral conduct, Leopold launched into yet another philosophic issue: "Was the earth made for man's use or has man merely the privilege of temporarily possessing an earth made for other and inscrutable purposes?" Because he recognized that most people were heir to the mechanistic, anthropocentric scientific tradition or—like his wife, a devout Roman Catholic—professed one of the anthropocentric religions, he decided not to dispute the point. But he couldn't resist an observation: "It just occurs to me, however, in answer to the scientists, that God started his show a good many million years before he had any men for audience—a sad waste of both actors and music—and in answer to both, that it is just barely possible that God himself likes to hear birds sing and see flowers grow." Even granting that the earth is for man, there was still a question: "What man?" Four cultures had flourished in the Southwest without degrading it. What would be said about the present one?

> If there be, indeed, a special nobility inherent in the human race—a special cosmic value, distinctive from and superior to all other life—by what token shall it be manifest? By a society decently respectful of its own and all other life, capable of inhabiting the earth without defiling it? Or by a society like that of John Burroughs' potato bug, which exterminated the potato, and thereby exterminated itself?

We can only guess why Leopold decided not to publish "Some Fundamentals." One reason might have been a certain discomfort with the inconclusiveness of his philosophical arguments. Another might have had to do with criticism of his interpretation of the causes of soil erosion.

One colleague to whom he sent the draft for review, Morton M. Cheney of the lands division, took him to task for overemphasizing the destructiveness of erosion. Like most other foresters and scientists of the time, Cheney viewed erosion as a natural, ongoing geologic process, a "world-building factor" that would ultimately smooth the rough uplands of the Southwest and create an immense area of agricultural land. In retrospect, it is clear that Cheney was explaining to Leopold the classic geomorphological theory of landscape development of William Morris Davis, who described the stages through which mountains are uplifted and erode to peneplains. Davis's "cycles of erosion" had much in common with the prevailing model of forest ecology, the stages of plant succession to climax, which underlay forest policy in the Southwest. Both were developmental models, defining predictable stages leading to a stable endpoint or equilibrium that would persist indefinitely, unless assaulted by some force acting from outside the system. Thus Cheney articulated essentially the Forest Service prescription for management: protect the climax forests at the headwaters from fire, which will retard some erosion, and don't worry too much about erosion downstream, a natural process in any case.

Leopold didn't need a disquisition on erosion from Cheney. He had undoubtedly studied all that at Yale. But he knew it did not explain what he observed on the ground. He must also have been enormously frustrated by the unwillingness of his colleagues to recognize any responsibility on the part of the Forest Service to deal with the problem. One effect of Cheney's critique must have been to make Leopold realize that he was going to have to provide a much more persuasive analysis of the causes of erosion, starting from facts anyone could see. This he accomplished a year later in an extraordinary piece of observation and inferential reasoning dealing with the relationship between grazing, fire, plant succession, and erosion. "Grass, Brush, Timber and Fire in Southern Arizona" still stands as a landmark in ecological literature. In it he issued a direct challenge to Forest Service dogma: "Fifteen years of forest administration were based on an incorrect interpretation of ecological facts and were, therefore, in part misdirected" (Leopold 1924a).

Leopold's theory, in brief, held that Southwestern watersheds had maintained their integrity despite centuries of periodic wildfire set by lightning or Indians. overgrazing by cattle beginning in the 1880s thinned out the grass needed to carry fire, and brush species, now free of both root competition and fire damage, began to "take the country," thus further reducing the carrying capacity for cattle. By the time brush and, higher up, yellow pine had grown dense enough to carry fire, the Forest Service had arrived on the scene to Prevent it. Trampling by cattle along watercourses allowed erosion to start, and the grass was no longer sufficient to prevent devastation. Contrary to Forest Service doctrine, Leopold found evidence that fire, far from being an unmitigated evil, was natural and even beneficial, and that grass was a much better conserver of watersheds than were trees or brush. While the Forest Service was willing to acquiesce in some overgrazing and erosion in order to reduce the fire hazard, Leopold was willing to rake an added risk of fire in order to maintain the integrity of the watersheds. The Forest Service was thinking of the commodity values of cattle and timber. Leopold was thinking of the whole system.

Although he still had not broken completely out of the mold in which he had been educated and there was much research yet to come, especially on such matters as fire ecology, climate change, and the role of wildlife, Leopold finally had a theory to explain the degradation of Southwestern watersheds and was able to consider its implications for administrative policy and for human ethical behavior. Even as he was working out the new theory, he also prepared a comprehensive "Watershed Handbook" for the district, to train foresters to analyze and treat erosion problems, and he took his case for conservative land use to the public in an article, "Pioneers and Gullies," in *Sunset Magazine*. In addition to posing several short- and medium-range policy options—artificial works, better regulated grazing permits, a leasing system—he also addressed the long-range ethical issue, for the first time in print. But it was just a short prediction: "The day will come when the ownership of land will carry with it the obligation to so use and protect it with respect to erosion that it is not a menace to other landowners and to the public" (Leopold 1924b). Frustrated as he was by the failure of the Forest Service to address the erosion problem, Leopold realized that the issue ultimately would have to be joined on private lands. Hence the need for a sense of ethical obligation—supported, he hoped, by a legal concept of contingent possession or some other public policy.

It would be another decade before Leopold would publish anything substantial on conservation as a moral issue. But when he did, it would relate integrally to his deepened ecological understanding of Southwestern watersheds. Once he could explain in sufficient ecological detail the phenomenon of erosion, or other aspects of the dynamic functioning of ecosystems through time, he would no longer need to call on the authority and terminology of the philosophers. He could stand secure on his own ground. Was the earth living or not? Was the earth primarily for man's use or not? Definitive philosophical answers to such questions no longer seemed critical as the basis for an ethic. It was more important to grasp the interrelatedness of all elements of the system, physical and biological, natural and cultural, and to appreciate the extent to which human as well as biotic interests were dependent upon maintenance of the integrity of that system through time.

In 1924 Leopold left the Southwest for Madison, Wisconsin, to assume a new position as associate director of the Forest Products Laboratory. The laboratory at that time was the agency's principal research center. Though he was offered the position in part because of his intense scientific curiosity and interest in research as demonstrated in his work on Southwestern watersheds, the offer did not imply wholesale acceptance of his ideas within the Forest Service. Nor did the new job offer a fully satisfying scope for his interests. As its name implied, the laboratory dealt more with research concerning the uses of trees than with the living forest.

Four years later, Leopold left the Forest Service to venture fulltime into the relatively new field of game management, which had been his hobby long before it became his profession. Drawing on his experience inspecting and analyzing the condition of watersheds in Southwestern forests, he began a series of game surveys in the Northcentral states with funding from the Sporting Arms and Ammunition Manufacturers' Institute. The surveys augmented his understanding of interrelationships between wildlife and the land community and launched him on another major project, the writing of a comprehensive text for the new field. More or less unemployed during the darkest years of the depression, 1931 and 1932, while he was writing his now-classic *Game Management*, Leopold readily accepted a temporary position with Franklin Roosevelt's New Deal in the spring of 1933 and headed back to the Southwest. His assignment was to supervise erosion control work by the newly established Civilian Conservation Corps in dozens of camps in the national forests.

He found the problems as evident as when he had left the region a decade earlier. The causes and processes of erosion were still not well understood within the Forest Service, and there had as yet been no changes in grazing regulations. The worst problems were on private land, giving urgency to the need Leopold had identified in "Pioneers and Gullies" for institutional incentives to conservative land use. Small wonder then that he drew heavily on watershed problems in the Southwest when, for the first time in public or in print, he developed the case for a land ethic. The occasion was the annual John Wesley Powell Lecture to the Southwestern Division of the American Association for the Advancement of Science.

The Conservation Ethic

"The Conservation Ethic" was by all odds the most important address of Leopold's career. It begins with a passage familiar to readers of his later "Land Ethic": "When god-like Odysseus returned from the wars in Troy, he hanged all on one rope some dozen slave-girls of his household whom he suspected of misbehavior during his absence" (Leopold 1933). Disposal of property was a matter of expediency. Leopold went on to make the case for the extension of ethical criteria to all human beings, to social relationships, and eventually to the land community; he asserted that this extension of ethics was actually a process in ecological evolution. An ethic, he said, "may be regarded as a mode of guidance for meeting ecological situations so new or intricate, or involving such deferred reactions, that the path of social expediency is not discernible to the average individual." Discussing the role of ecology in history and its relationship to economics and eventually to ethics, he returned repeatedly to the vexing problem of erosion.

Leopold could identify three possible motives for soil conservation: self-interest, legislation, and ethics. Self-interest did not pay, especially on the marginal private land most subject to abuse. Legislative efforts to regulate land use would force tax delinquency, especially during a depression, and public ownership could not possibly go far enough. "By all the accepted tenets of current economics and science we ought to say, 'let her wash,'" Leopold concluded. Staple crops were overproduced, population was stabilizing, science was still raising yields, government was spending millions to retire lands, "and here is nature offering to do the same thing free of charge; why not let her do it?" This was economic reasoning. "yet no man has so spoken." To Leopold this fact was significant: "It means that the average citizen shares in some degree the intuitive and instantaneous contempt with which the conservationist would regard such an attitude." In this intuitive reaction Leopold saw the embryo of an ethic.

"The Conservation Ethic" is curiously transitional, much more economically based and management-oriented than his 1923 discussion of conservation as a moral issue and without appeals to Ouspensky or other philosophers. By the time the speech was published, Leopold had accepted a new chair of game management created for him in the Department of Agricultural Economics at the University of Wisconsin. There he would help devise new social and economic tools to deal with problems of land utilization and resource conservation. His emphasis in the address on ecologically based management by private owners on their own land was almost certainly in anticipation of his new position, where he would work not only with government agencies but also, and especially, with farmers. Even here, however, there is a mixture of older concepts with new: "It is no prediction but merely an assertion that the idea of controlled environment contains colors and brushes where- with society may someday paint a new and possibly a better picture of itself." The "idea of controlled environment"—this confidence in the possibility of control, which permeates "The Conservation Ethic" and also *Game Management*, pub- lished the same year, is straight out of the Progressive Era conservation tradition of Gifford Pinchot. It assumes that scientific intelligence can learn enough about the system to exert complete control, an assumption that Leopold's invocation of an ecological attitude was even then beginning to challenge. The resolution would await yet another stage in his evolution of a land ethic.

In his new position at the university, he applied his concern about watershed integrity and soil erosion to southwestern Wisconsin. In a severely eroded water- shed known as Coon Valley, Leopold and colleagues persuaded H.H. Bennett, chief of the federal Soil Erosion Service, to establish a pioneering demonstration on the integration of land uses—soil conservation, pasturage, crops, forestry, and wild- life—that would use trained technicians and hundreds of local farmers. Not long after the project began, Leopold expressed the difficulty and necessity of integrating land uses on private land in a wide-ranging critique of the single-track agencies and public-purchase panaceas of the New Deal. "Conservation Economics" is another classic Leopold essay, further substantiating the case for wise use of private as well as public land.

Philosophers speak of the need for a metaphysic, a theory of the scheme of things, as an undergirding for an ethic. Leopold no longer looked to philosophers for a metaphysic; he simply wanted to understand what was happening on the land. But he was still struggling to understand how the system functions and how human beings relate to it. In 1935 he returned again to the erosion puzzle in the Southwest, trying his hand at working out a theory that would be mutually acceptable to for- esters, ecologists, geologists, and engineers concerning the role of man, and other factors such as climate and topography, in causing erosion. In order to counter an explanation of climate-induced synchronous timing of erosion episodes proposed by the eminent Harvard geologist Kirk Bryan, Leopold wrote "The Erosion Cycle in the Southwest," in which he came up with an elaborate theory of random timing that allowed a role for human agency. As with his 1923 effort, he sent the manuscript for review, including a copy to Bryan. And again, after receiving Bryan's critique, he decided not to publish; instead, he urged his son Luna to study at Harvard with Bryan to see if he could solve the problem. (World War II intervened and Aldo died before Luna was able to complete his Ph.D. dissertation on "The Erosion Problem of Southwestern United States" at Harvard in 1950.)

Throughout the second half of the decade, Leopold continued his intellectual struggle for a better understanding of the system. He sensed the inadequacy of prevailing models without being able to put his finger on the problem. Sometimes, as ideas grow and change, a certain dramatic experience can trigger a rearrangement of elements, resulting in a new theory of the scheme of things. For Leopold the trigger might have been the juxtaposition of several key field experiences in the mid-1930s: a trip to study game management and forestry in Germany, during which he was appalled by the highly artificial system of management that created a host of unanticipated problems; the acquisition of his own sand country farm, where he began to experience first-hand the imponderables of even the best-intentioned management; and, perhaps most vital, two hunting trips to the Rio Gavilan in the Sierra Madre of northern Chihuahua.

The Sierra Madre—just south of the border from the Southwest Leopold had struggled so long to understand, but protected from overgrazing by Apache Indians, bandits, depression, and unstable administration—still retained the virgin stability of its soils and the integrity of its flora and fauna. The Gavilan River still ran clear between mossy, tree-lined banks. Fires burned periodically without any apparent damage, and deer thrived in the midst of their natural predators, wolves and mountain lions. "It was here," Leopold reflected years later, "that I first clearly realized that land is an organism, that all my life I had seen only sick land, whereas here was a biota still in perfect aboriginal health." The vital new idea for Leopold was the concept of biotic health. It was that idea that finally gave him a model, a way of conceptualizing the system, that could become the basis for his mature philosophy.

A Biotic View of Land

Leopold worked out his first comprehensive statement of the new scheme in a paper titled "A Biotic View of Land," which he delivered in 1939 to a joint session of the Ecological Society of America and the Society of American Foresters. The biotic idea represented a shift from the older conservation idea of economic biology, with its emphasis on sustained production of resources or commodities, to a recognition that true sustained yield requires preservation of the health of the entire system. Scarcely five years earlier, Leopold himself had asserted that "the production of a shootable surplus is the acid test of the sufficiency of a conservation system." Now he was distinguishing between the old economic biology, which conceived of the biota as a system of competitions in which managers sought to give a competitive advantage to useful species, and the new ecology, which "lifts the veil from a biota so complex, so conditioned by interwoven cooperations and competitions, that no man can say where utility begins or ends" (Leopold 1939). Though he had clearly been moving toward such a view years earlier in his work on Southwestern watersheds, the 1939 statement marked the first significant publication in which his thinking about wildlife was fully integrated in the new conception. Thus, better than anything else he wrote, "A Biotic View of Land" signaled the maturity of Leopold's thinking.

It was here that he first presented the image of land as an energy circuit, a biotic pyramid: "a fountain of energy flowing through a circuit of soils, plants, and animals." It was here too that he drew ecological interrelationships into an evolutionary context. The trend of evolution, he suggested, was to elaborate and diversify the biota, to add layer upon layer to the pyramid, link after link to the food chains of which it was composed. Leopold posited a relationship between the complex structure of the biota and the normal circulation of energy through it—between the evolution of ecological diversity and the capacity of the land system for readjustment or renewal, what he would come to term land health.

Biotas seemed to differ in their capacity to sustain conversions to human occupancy. Drawing once again on his understanding of the Southwest, Leopold contrasted the resilient biota of Western Europe, which had maintained the fertility of its soils and its capacity to adapt to alterations despite centuries of strain, with the semiarid regions of America, where the soil could no longer support a complex pyramid and a "cumulative process of wastage" had set in. The organism would recover, he explained, "but at a low level of complexity and human habitability." Hence his general deduction: "the less violent the man-made changes, the greater the probability of successful readjustment in the pyramid."

The biotic idea Leopold articulated, though deeply a product of his own thought and experience, was part of a larger conceptual reorientation in the biological sciences during the 1930s. These were the years, according to historians of science, when various strands of evolutionary and ecological theory, separated during the furor over Darwin's *Origin of Species* (1859), began to fuse into a broad, unified theory. Among ecologists it has become known as the "ecosystem" concept, after a term suggested by the British ecologist A.G. Tansley in 1935. (Leopold's biotic idea, with its image of the land pyramid, probably owes more to another British ecologist, Charles Elton, and to an American scientist, Walter P. Taylor, both personal friends. Leopold did not use the term "ecosystem" in his writing, but he used "ecology" or "biotic idea" to express the same concept.) The new conception postulated a single integrated system of material and energy. The system was not driven by any one factor, such as climate, and hence was not best understood as developing through predictable stages to a stable endpoint, as in the older models of forest succession and erosion that had troubled Leopold in the Southwest. Rather, it was in constant flux, subject to unpredictable perturbations that entailed a continual process of reciprocal action and adjustment. Agents of disturbance— whether fire, disease, grazing animals, predators, or man—which had been viewed by early forest ecologists as acting from outside the system to subvert the normal successional stages or the climax equilibrium, were now viewed as functional components within the system.

Not everyone who adopted the ecosystem model drew from it the same implications for land management or for ethical behavior. Human beings, an exceptionally powerful biotic factor, were now clearly located *within* the system; did that mean they could manipulate it to their own ends with moral impunity? Donald Worster, who has written about the history of ecological thought in *Nature's Economy*, has noted the emphasis in recent ecosystem analysis on concepts such as productivity, biomass, input, output, and efficiency, which he views as metaphors for the modern corporate, industrial system bent on economic optimization and control. He also views modern ecosystem analysis as reductionist, breaking down the living world into readily measurable components that retain no image of organism or community. He is, therefore, highly skeptical of the ecosystem concept as the basis for an environmental ethic (Worster 1977).

A caveat is in order. Science alone is hardly an adequate basis for an ethic, and Leopold realized as much. For all his commitment to understanding how the system functioned, even Leopold did not insist that science provided all the answers. What he gained from the biotic view was a new humility about the possibility of ever understanding the system fully enough to exercise complete control. As he expressed it in a speech on "Means and Ends in Wildlife Management," some managers had admitted their "inability to replace natural equilibria with artificial ones," and their unwillingness to do so even if they could. The objective of management, as he now viewed it, was to preserve or restore the capacity of the system for sustained functioning and self-renewal. He would do this by encouraging the greatest possible diversity and structural complexity and minimizing the violence of man-made changes. The techniques of management might remain much the same, but the ends (of wildlife management, at least) were now fundamentally altered. The ends, he realized, were a product of the heart as much as of the mind.

Leopold came to a deeper personal understanding and appreciation of the new biotic idea and its implications for land management through his own participation in the land community at his "shack" in central Wisconsin's cutover, plowed up, worn out, and eroded sand country. All during the late 1930s and into the 1940s, while he was struggling to put the biotic idea and the land health concept on paper, he was also struggling to rebuild a diverse, healthy, esthetically satisfying biota on his farm. His journals of the shack experience record a daily routine of planting and transplanting—wildflowers, prairie grasses, shrubs and trees, virtually every species known to be native to the area. But the journals also record his tribulations. Take pines, for example, of which the family planted thousands every year. The first year, 1936, more than 95 percent were killed by drought within three months. Another year, rabbits attracted to brush shelters he had built for the birds trimmed three-quarters of the white pines in the vicinity. Other times the culprits were deer or rust or weevils or birds alighting on the candles or vandals cutting off the leaders or flood or fire. Fire could be discouraging, especially if set by a trespassing hunter, but it also brought new life—sumac and wild plum, blackberry, bluestem, poison ivy and, most exciting to Leopold, natural reproduction of jackpines from cones undoubtedly opened by the heat.

The shack experience engendered in Leopold a profound humility in his use of the manager's tools, as he became acutely aware of the innumerable, ofttimes inscrutable factors involved in life and death, growth and decay. It also led him to ponder the basis for the individual decisions he found himself making every day—whether to plant something as useless as a tamarack (yes, because it was nearly extinct in the area and it would sour the soil for lady's-slippers), what to do about the sandblow on the hill (leave it as testimony to history and also as a habitat for certain species like little Linaria that would grow only there), whether to favor the birch or the pine where the two were crowding each other (he loved all trees, but he was "in love" with pines). He realized that ethical and esthetic values could be a guide for individual decisions, not a substitute for them. And he also gained a sense of belonging to something greater than himself, a continuity with all life through time. This he expressed in a series of vignettes that ultimately found their way into *A Sand County Almanac*.

Toward a Land Ethic

Toward the end of the 1940s, Leopold tried again to make the case for a conservation ethic. This new effort, for a 1947 address to the Garden Club of America on "The Ecological Conscience," was less ambitious than his 1933 "Conservation Ethic." He drew on four issues in which he had been involved in Wisconsin, including a wrenching debate about the state's "excess deer" problem that had preoccupied him for years. The public thought only about conserving deer because they were unable to see the land as a whole. As in each of his previous efforts to make the case for a sense of obligation to the community going beyond economic self-interest, he also addressed the problem of soil erosion, this time analyzing the failure of soil conservation districts in Wisconsin to achieve anything beyond those few reme-dial practices that were immediately profitable to the individual farmer. From his unpublished papers we know that he had been trying for several years to articulate the land health concept in relation to land use and erosion, but he made no effort to do so in "The Ecological Conscience" (Leopold 1947). The speech was not particu-larly significant—except for one pregnant sentence setting forth the criteria of an ethic, which would ultimately find a more appropriate context in his most famous essay, "The Land Ethic."

Later that year and in early 1948, Leopold substantially reshaped and revised the collection of essays for which he had been seeking a publisher all decade. It would begin with a selection of vignettes from the shack, arranged by month in the style of an almanac. That would be followed by sketches recounting various episodes in his career that taught him the meaning of conservation. Then he would conclude with several meatier essays, culminating in a comprehensive statement of his ethical philosophy. "The Land Ethic" incorporates segments from three previous essays, all thoroughly reworked and integrated with new material to reflect his current thinking. From "The Conservation Ethic" he drew the notion of the ecological and social evolution of ethics and the role of ecology in history; from "A Biotic View of Land" the image of the land pyramid, of land as an energy circuit; and from "The Ecological Conscience" the case for obligations to land going beyond economic self-interest. His efforts of a decade to articulate the concept of land health and the relationships between economics, esthetics, and ethics, filling numerous handwrit-ten, heavily interlined pages, finally found compelling expression. And the whole came to a focus in the most widely quoted lines in the entire Leopold corpus:

"A thing is right when it tends to preserve the integrity, stability, and beauty of the biotic community. It is wrong when it tends otherwise" (Leopold 1949).

Integrity, stability, beauty: the fundamental criteria of the land ethic. *Integrity*, referring to the wholeness or diversity of the community: the precept to retain or restore, insofar as possible, all species still extant that evolved together in a particular biota. *Stability*, embodying the concept of land health: the precept to maintain or restore an adequately complex structure in the biotic pyramid, so that the community has the capacity for sustained functioning and self-renewal. *Beauty*, the motive power of the ethic: the precept to manage for values going beyond the merely economic—and, probably also, an allowance for the subjective tastes of the individual. The three tenets were interrelated. Elsewhere in his writings Leopold had referred to an assumed relationship between the diversity and stability of the biotic community as "the tacit evidence of evolution" and "an act of faith." And as early as 1938 he had posited a relationship among all three tenets, and utility as well, in an unpublished fragment titled "Economics, Philosophy and Land": "We may postulate that the most complex biota is the most beautiful. I think there is much evidence that it is also the most useful. Certainly it is the most permanent, i.e., durable. Hence there is little or no distinction between esthetics and utility in respect of biotic objective."

The three cardinal tenets of the land ethic, first voiced in his address "The Ecological Conscience," can also be discerned, in somewhat different terminology, in Leopold's earlier formulations of his ethical philosophy in 1923 and 1933. This supports a conclusion that his mature expression involved a deeper understanding of the functioning of the land system and a more cogent articulation rather than a change in fundamental values. All these formulations, and "A Biotic View of Land" as well, were premised on a conception of land as an interrelated, indivisible whole, a system that deserved respect as a whole as well as in its parts. In 1923 Leopold drew on his intuitive perceptions, buttressed by the concepts and terminology of Ouspensky and other philosophers and poets. Later he would base his conception of the land community on the findings of ecology, but his willingness to trust to intuition remained.

In 1933, for example, he wrote, "Ethics are possibly a kind of advanced social instinct in-the-making," revising the phrase in 1948 to "community instinct." Each formulation also emphasized the notion of *obligation* to the whole, rather than focusing on the *rights* of individual constituents, whether human, animal, vegetable, or mineral. Leopold did not deny that nonhuman entities had rights, and occasionally he even referred to a species' "biotic right" to existence, but he was too much concerned with securing acceptance of his major premises to risk alienating people by entering the thicket of the rights debate. (Because most philosophers are concerned with individuals rather than with communities and with rights rather than with obligations, Leopold's land ethic has often been distorted, disparaged, or dismissed in philosophical circles. Among those who have studied his work carefully enough to puzzle through some of the vagaries of language and seeming inconsistencies, Bryan Norton places Leopold in the tradition of American pragmatism and J. Baird Callicott identifies him as heir to the bio-social ethical tradition of David Hume and Charles Darwin. Both are reasonable in light of Leopold's education, reading, and experience.)

Leopold articulated his ethical philosophy out of a profound conviction of the need for moral obligation in dealing with the dissolution of watersheds in the Southwest. Hence his emphasis on the integrity of the system. Each successive reformulation of his philosophy was stimulated at least in part by his continuing concern for the erosion problem and advances in his understanding of the ecological processes involved. Remarkably, it was his compelling concern and curiosity about the phenomenon of erosion, which was never a major professional responsibility, rather than his lifelong interest in wildlife, which became his profession, that led him to his conviction of the need for a land ethic and his understanding of the biotic idea on which it was grounded. But once he had grasped the biotic concept, through which he finally integrated wildlife fully into his understanding of the functioning of the land community, it was his sensitivity to the esthetics of wildlife that would enable him to convey a sense of the land community and the land ethic to others. *A Sand County Almanac* is the case in point.

Leopold's fascination with the new biotic concept, especially the role of evolution, energy, and land health, led to an explosion of the classic literary essays for which he is best known today: "Marshland Elegy," with its haunting image of sandhill cranes, evolutionarily among the most ancient of species, standing in the peat bogs of central Wisconsin "on the sodden pages of their own history"; "Clandeboye," where the western grebe, also of ancient lineage, "wields the baton for the whole biota"; "Odyssey," the saga of two atoms cycling through healthy and abused systems; "Song of the Gavilan," where food is the continuum in the stream of life; "Guacamaja," a disquisition on the physics of beauty, recording the discovery of the numenon of the Sierra Madre, the thick-billed parrot; and "Thinking Like a Mountain," in which the wolf becomes metaphor for the functioning system.

These essays had a purpose with respect to Leopold's notion of the evolution of an ethic. Their purpose was to inspire respect and love for the land community, grounded in an understanding of its ecological functioning. Leopold would motivate that understanding of the whole by focusing the reader's attention on the subtle dramas inherent in the roles of wolf, crane, grebe, parrot, even atom, in the scheme of things. The essays were Leopold's attempt to develop a metaphysic, or an esthetic—to stimulate perception that might lead people to the transformation of values required for a land ethic. He would motivate not by inciting fear of ecological catastrophe or indignation about abused watersheds but rather by leading people from esthetic appreciation through ecological understanding to love and respect.

He had thus come full circle in his own development—from his youth, in which esthetic appreciation for wildlife provided his personal motivation to enter a career in conservation, through his professional experience in forestry and his concern about watersheds, which stimulated his consciousness of the need for an ethical obligation to land, to his maturity as an ecologist, when he successfully integrated all the strands of his previous experience. Reflecting on the process by which he himself had come to ecological and ethical consciousness, he would now inspire others along a similar route.

Acknowledgements

Biographical details and generalizations are based primarily on the Aldo Leopold papers in the University of Wisconsin Division of Archives and interviews with family members and friends of Aldo Leopold. All quotations from letters and unpublished manuscripts are from items in the Aldo Leopold papers.

References

Flader, Susan L. 1974. Thinking like a Mountain: Aldo Leopold and the Evolution of an Ecological Attitude toward Deer, Wolves, and Forests. Columbia, MO: University of Missouri Press.

Hays, Samuel P. 1959. Conservation and the Gospel of Efficiency: The Progressive Conservation Movement, 1890-1920. Cambridge, MA: Harvard University Press.

Leopold, Aldo. 1911. The busy season. The Carson [National Forest] Pine Cone (July).

Leopold, Aldo. 1924a. Grass, brush, timber, and fire in southern Arizona. Journal of Forestry 22(October):1-10.

Leopold, Aldo. 1924b. Pioneers and gullies. Sunset Magazine 52(May): 15-16, 91-95.

Leopold, Aldo. 1933. The conservation ethic. Journal of Forestry 31(October):634-643.

Leopold, Aldo. 1939. A biotic view of land. Journal of Forestry 37(September):727-730.

Leopold, Aldo. 1947. The ecological conscience. The Bulletin of the Garden Club of America (September):45-53.

Leopold, Aldo. 1949: A Sand County Almanac and Sketches Here and There. New York: Oxford University Press.

Leopold, Aldo. 1979. Some fundamentals of conservation in the Southwest. Environmental Ethics 1(Summer):131-141.

Worster, Donald. 1977. Nature's Economy: The Roots of Ecology. San Francisco: Sierra Club Books.

ORIGINAL PUBLICATION DETAILS

By Susan Flader, Professor of History, University of Missouri, Columbia, Missouri.

Flader, Susan. 1987. Aldo Leopold and the evolution of a land ethic. *In* Aldo Leopold: The Man and His Legacy, ed. Thomas Tanner, 3-24. Ankeny, IA: Soil and Water Conservation Society.

The Farmer as Conservationist:
Aldo Leopold on Agriculture

Curt Meine

In February 1939, as part of the Wisconsin Farm and Home Week observance at the University of Wisconsin, Aldo Leopold presented an address entitled "The Farmer as a Conservationist" (Leopold 1939). Leopold began his remarks with these words:

"When the land does well for its owner, and the owner does well by his land—when both end up better by reason of their partnership—then we have conservation. When one or the other grows poorer, either in substance, or in character, or in responsiveness to sun, wind, and rain, then we have something else, and it is something we do not like.

"Let's admit at the outset that harmony between man and land, like harmony between neighbors, is an ideal-and one we shall never attain. Only glib and ignorant men, unable to feel the mighty currents of history, unable to see the incredible complexity of agriculture itself, can promise any early attainment of that ideal. But any man who respects himself and his land can try to" (Leopold 1939).

This quotation is vintage Leopold, displaying his characteristic mix of idealism and practicality, expressing his dual concern for the fate of man and land. It was, in fact, the first time in print that he gave his classic definition of conservation as the state of "harmony between man and land." In it too we see what Leopold had learned over a period of many years: that when one addresses the subject of agriculture, one takes on a subject of immense proportions.

For those who know of Leopold as the poet of *A Sand County Almanac*, or as an early voice for wilderness preservation, or as a founding father of wildlife management, it may come as a surprise to know that Leopold, while not a farmer himself, did work on a number of agricultural fronts. It is one of the less heralded aspects of his multifaceted career, but one that is bound to become increasingly important in these times of transition on the rural landscape.

Even a cursory review of Leopold's career shows that he was involved in agriculture throughout his professional life. He spent 19 years with the USDA's Forest Service. As a pioneering formulator and practitioner of game management, he worked closely with farmers and became an expert observer of the farm landscape at a time—the 1920s and 1930s when, like today, that landscape was undergoing great change. As a professor, Leopold taught for 15 years in one of the nation's top colleges of agriculture. As a writer, he wrote for and about farmers extensively, as

34

much perhaps as on any topic. Finally, as a conservation philosopher, he made a special effort to define the role farming played in the greater equation of mankind's relationship to the natural environment.

The Farmer Must Do Conservation

The year was 1928. Leopold was the somewhat disgruntled associate director of the Forest Products Laboratory in Madison, Wisconsin. He had spent four years in the position, waiting for a promised promotion that never came. Through his work and writing, Leopold was already a respected figure in conservation. When word spread that he was looking for new work, opportunities quickly arose.

For more than 10 years Leopold had been devoting much of his spare time to game conservation. Game management, as we now know it, existed only in its embryonic stages. For years, Leopold had been promoting the idea, a new idea, that wild game could be raised on a sustained-yield basis, much as foresters raised trees. Moreover, the idea was not merely to rear game and then release it to be shot, but to manipulate habitat so that, in effect, the game raised itself. This was a radical and unproven notion, but it was an important one—and growing more important with the passing seasons. Squeezed between a vastly increased hunting public and an increasingly intensified agriculture, game populations in the 1920s were plummeting. Action had to be taken if hunting, or even casual observation of game, were to remain a viable proposition.

On May 22, 1928, Leopold signed a contract with the Sporting Arms and Ammunition Manufacturers' Institute, a consortium of major firearms manufacturers, to conduct an unprecedented survey of game conditions across the country. So primitive was the state of wildlife science that it lacked even the most basic information on game ranges, life histories, food and habitat needs, population dynamics, and susceptibility to hunting pressure. The game survey was to make at least a start in gathering such information.

The game survey constituted a major landmark in Leopold's professional development. It was an opportunity to study what had been a life-long interest—in his own words, "to make my hobby my profession." Leopold was already an astute observer of land; the game survey would hone his talent into genius. The method was straightforward: Leopold spent a month or so in each state, meeting its local experts, learning its geography, touring its backroads, talking to an amazing assortment of sportsmen, administrators, botanists, zoologists, farmers, professors, wardens, and foresters. The result, he hoped, would be a fair estimate of a state's game resources and a growing body of knowledge about game biology.

That summer of 1928, Leopold completed his first tentative surveys in Michigan, Minnesota, and Iowa. Conditions, of course, varied according to species and locality, but after his first months on the job, Leopold had begun to find evidence to support the one overriding suspicion of the times: that the sudden intensification of agriculture was eliminating the food and cover plants required by the majority of game species. Fencerows, borders, woodlots, remnant prairies, and wetlands were disappearing from the Midwestern farmscape, and the quail, prairie chicken, grouse, snipe, woodcock, and in some areas even rabbits and squirrels were disappearing with them. This realization came as no surprise, but Leopold, for the first time, was giving it factual substance.

By 1929 it became apparent that the game survey as originally designed was too ambitious. Leopold and his sponsors decided to confine its coverage, at least for the time being, to the north-central block of states: Minnesota, Wisconsin, Michigan, Ohio, Indiana, Illinois, Iowa, and Missouri. Leopold spent much of the next year and a half on the road, crisscrossing the Midwest, coming to know its contours with an intimacy that only grew with each new effort.

It was at this point that Leopold first began to devote his attention to the question of the farmer's role in wildlife conservation. The question had arisen before in Leopold's work and writing, but not with the same urgency. His answer was unequivocal:

"Most of what needs doing must be done by the farmer himself. There is no conceivable way by which the general public can legislate crabapples, or grape tangles, or plum thickets to grow up on these barren fencerows, roadsides, and slopes, nor will the resolutions or prayers of the city change the depth of next winter's snow nor cause cornshocks to be left in the fields to feed the birds. All the nonfarming public can do is to provide information and build incentives to which farmers may act" (Leopold 1933).

And those were the keys: to provide information and build incentives. Farmers had no more idea about the needs of game animals than anyone else, so Leopold began to write his earliest articles for farmers on the subject. The first apparently was a 1929 article, "How the Country Boy or Girl Can Grow Quail."

The second point—building incentives—proved more provocative. Throughout the 1920s, farmers were increasingly posting their lands against hunting, in order to keep their undisciplined city cousins out of their fields. The last thing farmers wanted was more game. Posting became so widespread that conscientious sportsmen were forced to come up with alternative proposals.

In 1929 and 1930, Leopold took on this issue in his work as chairman of the Game Policy Committee for the American Game Conference. The purpose of the committee, which included many of the nation's foremost game experts, was to draw up a definitive national game policy, a statement that was destined to guide the wildlife profession for the next 40 years. The policy, most of which Leopold himself wrote, was premised on the idea that "only the landowner can practice management efficiently, because he is the only person who resides on the land and has complete authority over it." A principal recommendation of the policy read:

"Recognize the landowner as the custodian of public game on all other land, protect him from the irresponsible shooter, and compensate him for putting his land in productive condition. Compensate him either publicly or privately, with either cash, service, or protection, for the use of his land and labor, on condition that he preserves the game seed and otherwise safeguards the public interest. In short, make game management a partnership enterprise to which the landowner, the sportsman, and the public each contribute appropriate services, and from which each derives appropriate support" (Leopold et al. 1931).

The important point here again is Leopold's steadfast conviction that the farmer, for reasons both practical and philosophical, was the one to do conservation. At the time, Leopold was speaking only of the conservation of game animals, but in the important years yet to come he would extend this notion to include nongame wildlife, plants, soil, water, and even scenic values. And it was this emphasis on individual landowner action that would lead him to be such an outspoken critic of the New Deal's top-heavy approach to conservation.

Building Communication Channels

A great deal of Leopold's success as a conservation leader must be attributed to his unique communication skills. This was never so true as when he was working with farmers, whether in print, in the classroom, over the airwaves, or in personal contacts. This skill undoubtedly derived from his curiosity, as infectious as it was insatiable, about the land itself—its human and nonhuman denizens, its dynamic processes, its history and destiny. Many were the farmers who themselves learned to see their land more acutely as a result of Leopold's insight.

In his days as a forester, Leopold had gained a solid appreciation of rural psychology. Before getting down to business he enjoyed talking over crop prospects, soils, local lore, the vicissitudes of the weather and seasons. One day in the summer of 1931, Leopold was driving through Dane County, west of Madison, scouting potential hunting grounds for the upcoming season. Upon coming to a mail stop on the Chicago and Northwestern line known as Riley, he pulled in at a farm for a drink of water. He and the farmer, a man named Reuben Paulson, talked over their mutual concerns. Paulson needed relief from poachers and trespassers. Leopold needed a place to try out his ideas about game management. Paulson organized 11 of his neighboring farmers, while Leopold called on four of his hunting partners from town. Thus was born the Riley Game Cooperative, an important example of the kind of cooperative management arrangement that the American Game Policy had tied to promote. Riley became a significant center for Leopold's work (as well as his recreation) for years to come, and the Riley farmers became his close friends. Leopold was, in turn, a trusted advisor to them (Leopold and Paulson 1934).

Riley was but one of several cooperative arrangements with farmers that Leopold helped organize and develop in the 1930s and 1940s. After Leopold joined the University of Wisconsin, these farms played an important role as study areas for the first generation of professionally trained wildlife managers. Much of the groundbreaking research in wildlife management was in fact performed on farms in southern Wisconsin. The importance of Leopold's style in these matters cannot be overstated. Even as his ecological vision sharpened in the 1930s and his conservation message became correspondingly more impassioned, Leopold never forgot that, in the Midwestern farmbelt at least, it was the farmer who was on the front lines of conservation and had to be treated accordingly.

Leopold's appointment to the chair of game management at the University of Wisconsin in 1933 provided him, for the first time, a secure position from which to implement his management ideas. It is important to note that the chair was initially established within the Department of Agricultural Economics, and from this point on in Leopold's work one sees an increasing sophistication in his views on rural culture. The department was the first of its kind in the nation, and the pioneering work on rural economics, particularly that performed by his good friend George Wehrwein, would have a lasting impact on his own approaches to land use reform.

Leopold's new position entailed a number of responsibilities, including acting as a wildlife extension specialist. It was in this capacity that he served as advisor to the Coon Valley soil conservation project, the world's first watershed-wide soil erosion control effort. Leopold's interest in soil erosion dated back to the early 1920s, when as a forest inspector on the national forests in the Southwest he initiated a remarkable personal study of the ecological cause-and-effect of soil erosion on the Southwestern range. With his work at Coon Valley, that interest reemerged in the Midwest, never again to go into eclipse. Situated in the erosion-prone driftless area of western Wisconsin, Coon Valley was in 1933 a wasted watershed, ruined by deforestation, poor tillage practices, overgrazing, and soil depletion. It was, in Leopold's bitter phraseology, "one of the thousand farm communities which, through the abuse of its originally rich soil, has not only filled the national dinner pail, but has created the Mississippi flood problem, the navigation problem, the overproduction problem, and the problem of its own future continuity" (Leopold 1935). The work of the new Soil Erosion Service—later renamed the Soil Conservation Service—would turn the situation around through a unique program of integrated land use. In an article describing the success of Coon Valley, Leopold later wrote:

"There are two ways to apply conservation to land.

"One is to superimpose some particular practice upon the preexisting system of land use, without regard to how it fits or what it does to or for other interests involved.

"The other is to reorganize and gear up the farming, forestry, game cropping, erosion control, scenery, or whatever values may be involved so that they collectively comprise a harmonious balanced system of land use.... The crux of the land problem is to show that integrated use is possible on private farms, and that such integration is mutually advantageous to both the owner and the public" (Leopold 1935).

Among his other extension activities as a professor of game management, Leopold instituted a short course for young farmers and presented a number of radio talks for farmers over WHA, the university-sponsored radio station. In both cases Leopold was trying to disseminate basic information on wildlife conservation to farmers. His first radio talk, delivered just after joining the university, was called "Building a Wisconsin Game Crop: Leaving Food and Cover." Others were similar: "The Farm Woodlot and the Bird Crop," "Game on the Modern Farm," and so on. The following excerpt gives the flavor of these talks:

"There are many little tricks for increasing the service of woods and vegetation to wildlife. Take the grapevine, for example. A new grape-tangle on or near the ground is usually good for a new covey of quail, provided there be food nearby. How to get a new grape-tangle quickly? Select a tree with a grapevine in its top. Cut the tree but not the vine, and let it lie. In one season the vine will weave an 'umbrella' over the down top which is hawk-proof and nearly manproof—a mighty fortress for bobwhite in even the deepest of snows. Leave a few cornshocks in the adjoining field and you have the 'makings' of a new covey range which your friends the quail will not long overlook. One of the real mysteries of nature is the promptness with which habitable niches in the cold wall of the world are filled by living things. Our own place in the scheme of things is not the less tolerable for making room for a few of our fellow-creatures....

"Your woodlot is, in fact, an historical document which faithfully records your personal philosophy. Let it tell a story of tolerance toward living things, and of skill in the greatest of all arts: how to use the earth without making it ugly."

Leopold would continue to provide this sort of encouragement throughout his university years. In 1938 he began writing similar pieces for the *Wisconsin Agriculturalist and Farmer* on a variety of topics: "Plant Evergreens for Bird Shelter," "Wild Flower Corners," "Look for Bird Bands," "Windbreaks and Wildlife." A few of these, in revised form, were incorporated into *A Sand County Almanac*.

Ecology and Agriculture

When the Dust Bowl of the mid-thirties hit, Leopold was already well on his way to the fully developed ecological philosophy that would mark his mature writings in *A Sand County Almanac*. The Dust Bowl only hastened this development and led directly to the question of what ecology had to offer by way of advice to agriculture in America. Increasingly, that advice would be framed in terms of what he called "land health": the ability of land as an integrated whole to regenerate itself. This was an issue inclusive of, but far greater than, agriculture alone.

The Dust Bowl was the upshot of the indiscriminate agricultural expansion of the post–World War I era. For Leopold and others it focused attention on the over-arching problem of how, in Leopold's tight phrasing, to "adjust men and machines to land." On April 15, 1935, coincidentally the day after the greatest dust storm yet swept out of the southern High Plains, Leopold delivered an address that he called "Land Pathology." In that unpublished speech he stated:

"This paper proceeds on two assumptions. The first is that there is only one soil, one flora, one fauna, and hence only one conservation problem. Each acre should produce what it is good for, and no two are alike. Hence a certain acre may serve one, or several, or all of the conservation groups. The second [assumption] is that economic and aesthetic land uses can and must be integrated, usually on the same acre. The ultimate issue is whether good taste and technical skill can both exist in the same landowner. This is a challenge to agricultural education."

After tracing the history of destructive land use in America, Leopold in the speech asked what might be done in the social and physical sciences to hasten "the needed adjustment between society as now equipped, and land use as now practiced." The profit motive, for a number of reasons, was insufficient. Public ownership was, to a true conservative like Leopold, a last resort and impractical to boot. Legislative compulsion was unpalatable. Besides, science by this time had "shown good land use to require much positive skill as well as negative abstention." The only alternative was a kind of land ethic, and this 1935 paper was one of Leopold's important early expressions of this maturing idea. He wrote:

"I plead for positive and substantial public encouragement, economic and moral, for the landowner who conserves the public values—economic or aesthetic—of which he is the custodian. The search for practicable vehicles to carry that encouragement is a research problem, and I think a soluble one. A solution apparently calls for a synthesis of biological, legal, and economic skills, or, if you will a social application of the physical sciences....

"I might say, defensively, that such a vehicle would not necessarily imply regimentation of private land use. The private owner would still decide what to use his land for; the public would decide merely whether the net result is good or bad for its stake in his holdings.

"Those charged with the search for such a vehicle must first seek to intellectually encompass the whole situation. It may mean something far more profound than I have foreseen."

The Dust Bowl was but one highly visible (and breathable) reminder that this sort of ethic was not merely a nice idea, but a necessary development. Leopold held that the improvement of economic tools had "exceeded the speed, or degree, within which it was good. Equipped with this excess of tools, society has developed an unstable adjustment to its environment, from which both must eventually suffer damage or even ruin. Regarding society and land collectively as an organism, that organism has suddenly developed pathological symptoms, i.e., self-accelerating rather than self-compensating departures from normal functioning. Granted that science can invent more and more tools, which might be capable of squeezing a living even out of a ruined countryside, yet who wants to be a cell in that kind of body politic? I for one do not."

Through the latter half of the 1930s, Leopold would devote increasing amounts of his time to defining the characteristics of healthy land and tracing the implications of that definition for land use. I will refrain from discussing the ecological implications of modern agricultural systems; others have treated this subject more ably and completely than I can here. Suffice to say that, after the experiences of the 1930s, agriculture would begin to come under the scrutiny of this new science of ecology, and Leopold himself would begin to apply the precepts of ecology more stringently in his analyses. Those precepts had biological as well as social implications. On both grounds, for example, he decried in an unpublished manuscript the trend toward monotypes, warning that "the doctrine of private profit and public subsidy pushes constantly toward an extreme degree of crop-specialization, toward the grouping of uses in large solid blocks. The idea of self-sufficient farm units is submerged. The interspersion of wild and tame crops approaches zero ...

[producing] a landscape just as monotonous as the inherent variability of soil will permit."

By this time, of course, Leopold had himself become the owner of a worn-out farm, and he and his family had begun the process of bringing it back to life. He did not consider himself a farmer, but there is an unmistakable sense of pride in husbandry that enters his writings from this point forward, a quality evident to anyone who has read *A Sand County Almanac*.

Perhaps the finest example of this, and one most salient to this discussion, is the article to which I referred earlier, "The Farmer as a Conservationist." It is one of Leopold's forgotten masterpieces—poignant and pointed, written in a delightful manner, and as pertinent today as it was 50 years ago.

The heart of Leopold's argument was that utilitarian motives had dominated the development of our agricultural system, to the general disadvantage of land, landowner, society, and even productivity itself. The neglect of the aesthetic qualities of land, while sounding abstract, had actually had very practical effects on the way people live on land. Again, Leopold speaks best for himself:

"If this were Germany, or Denmark, with many people and little land, it might be idle to dream about land use luxuries for every farm family that needs them. But we have excess plowland; our conviction of this is so unanimous that we spend a billion out of the public chest to retire the surplus from cultivation. In the face of such as excess, can any reasonable man claim that economics prevents us from getting a life, as well as a livelihood, from our acres?

"Sometimes I think that ideas, like men, can become dictators. We Americans have so far escaped regimentation by our rulers, but have we escaped regimentation by our own ideas? I doubt if there exists today a more complete regimentation of the human mind than that accomplished by our self-imposed doctrine of ruthless utilitarianism. The saving grace of democracy is that we fastened this yoke on our own necks, and we can cast it off when we want to, without severing the neck. Conservation is perhaps one of the many squirmings which foreshadow this act of self-liberation.

"One of the self-imposed yokes we are casting off is the false idea that farm life is dull? What is the meaning of John Steuart Curry, Grant Wood, Thomas Benton? They are showing us drama in the red barn, the stark silo, the team heaving over the hill, the country store, black against sunset. All I am saying is that there is also drama in every bush, if you can see it. When enough men know this, we need fear no indifference to the welfare of bushes, or birds, or soil, or trees. We shall then have no need of the word conservation, for we shall have the thing itself" (Leopold 1935).

Leopold's ideas on conservation, culture, and democracy were never so interwoven as when he addressed the topic of agriculture in American life. The Jeffersonian notion of a stable agrarian democracy of yeoman farmers had been left in the wake of the industrial revolution, but it was a buoyant ideal, and it resurfaced in Leopold's words. There was an important difference: where Jefferson had drawn his vision from his hopes for a healthy and lasting democratic republic, Leopold had had the benefit of 150 years of history and scientific advance, and his vision—deepening even as he wrote—was focused less on the policy than on the biology of healthy land. Yet, even through the intervening century and a half, the heart of the ideal remained. Freedom and individuality were still the points at issue.

"The landscape of any farm," Leopold wrote, "is the farmer's portrait of himself. Conservation implies self-expression in that landscape, rather than blind compliance with economic dogma" (Leopold 1939).

This notion of self-expression in the farm landscape was fundamental to Leopold's thinking. He wrote in personal notes at the time, in another context:

"I expect, and hope for, a wide range of individualism as the ultimate development of the wildlife idea. There are, and should be, farmers not at all interested in shooting, but keen on forestry, or wildflowers, or birds in general. There are, and should be, farmers keen about none of these, but hipped on coons and coon dogs. The more varied the media of individual expression, the more the collective total will add to [the] satisfaction of farm life."

That, in the end, was the focus of Leopold's work: the quality and satisfaction of farm life. To Leopold's thinking a farmscape stripped of all but its human economic components was not only at agronomic risk, but it was a waste of cultural potential. Conservation, conversely, sought to balance utility and beauty on the land; it was a challenge to use the earth without making it ugly.

A final quotation of Leopold's from a 1945 paper, "The Outlook for Farm Wildlife," speaks most directly to our farm situation today. Leopold concluded a review of trends in the farm wildlife situation by stating:

"In short, we face not only an unfavorable balance between loss and gain in habitat, but an accelerating disorganization of those unknown controls which stabilize the flora and fauna, and which, in conjunction with stable soil and a normal regimen of water, constitute land-health.

"Behind both of these trends in the physical status of the landscape lies an unresolved contest between two opposing philosophies of farm life. I suppose these have to be labeled for handy reference, although I distrust labels:

"1. *The farm is a food-factory*, and the criterion of its success is saleable products.

"2. *The farm is a place to live*. The criterion of success is a harmonious balance between plants, animals, and people; between the domestic and the wild; between utility and beauty.

"Wildlife has no place in the food-factory farm, except as the accidental relic of pioneer days. The trend of the landscape is toward a monotype, in which only the least exacting wildlife species can exist.

"On the other hand, wildlife is an integral part of the farm-as-a-place-to-live. While it must be subordinated to economic needs, there is a deliberate effort to keep as rich a flora and fauna as possible, because it is 'nice to have around.'

"It was inevitable and no doubt desirable that the tremendous momentum of industrialization should have spread to farm life. It is clear to me, however, that it has overshot the mark, in the sense that it is generating new insecurities, economic and ecological, in place of those it was meant to abolish. In its extreme form, it is humanly desolate and economically unstable. These extremes will some day die of their own too much, not because they are bad for wildlife, but because they are bad for farmers" (Leopold 1945).

On Seeking Harmony

How do we assess Leopold's words? In the half century since he wrote, conservation has evolved into environmentalism, while farming has moved toward agribusiness. Yet one need not read far into Leopold to appreciate the timeliness—or, perhaps more accurately, the timelessness—of his thoughts. They remain relevant so long as people live on land and so long as the human instinct for stewardship endures. But more to the point: Do they speak to the issues of the day?

We are told today that the changes in farm tenure taking place across the continent, including and especially the foreclosure problem, represent the inexorable workings of economic trends and that the family farm has itself become an expendable commodity. We are advised to abandon the Jeffersonian view of the farmer as a sentimental holdover from a bygone era. We are asked to forget the truth, so eloquently expressed by Leopold in both word and deed, that the farm is more than a place to grow food, that farms also grow farmers, and families, and plants and animals, both wild and tame. We apply patchwork solutions to problems that have been a long time in the building and that can only be confronted by a view of history, ecology, and economics that is as wide-reaching as agriculture itself.

I think conservationists have much to offer as the nation debates these points and seeks new answers. Leopold himself was as sound a voice as one could find. He was not one to make sentimental references to "the Heartland." He did not hold a romantic image of the farmer, but realized that farmers are as diverse and independently minded as any group of individuals. But he also realized that, fundamentally, a balanced society must be built on a stable system of agriculture and that this in turn must be built on an appropriate attitude toward the land that sustains us all.

His thoughts present us with a challenge. To agricultural scientists, historians, and economists, he challenges us to explore the all-too-neglected territory where separate disciplines meet.

To farmers, conservationists, and environmentalists, he challenges us to work together more than we have, to compare our aims, and to appreciate that whatever differences may exist between us pale before the common dilemma we all face as lovers of the land.

And finally, to all of us as citizens in a democracy, Leopold challenges us to consider what sort of society we wish to build: one that strives to squeeze the land for all it is worth, or one that seeks today and tomorrow the elusive harmony between humankind and land that Leopold called conservation.

References

Leopold, Aldo, et al. 1931. Report to the National Game Conference on an American Game Policy. *In* Transactions of the 17th American Game Conference. Washington, DC: American Game Association.

Leopold, Aldo. 1933. Game cropping in southern Wisconsin. Our Native Landscape (December).

Leopold, Aldo, and Reuben Paulson. 1934. Helping ourselves: Being the adventures of a farmer and a sportsman who produced their own shooting ground. Field and Stream 39(August):32-33, 56.

Leopold, Aldo. 1935. Coon Valley: An adventure in cooperative conservation. American Forests 41(March):205-208.

Leopold, Aldo. 1939. The farmer as a conservationist. American Forests 45(June):294-299, 316, 332.

Leopold, Aldo. 1945. The outlook for farm wildlife. *In* Transactions of the 10th North American Wildlife Conference. Washington, DC: American Wildlife Institute.

ORIGINAL PUBLICATION DETAILS

By Curt Meine, Institute for Environmental Studies, University of Wisconsin, Madison.

Meine, Curt. 1987. The farmer as conservationist: Aldo Leopold on agriculture. *Journal of Soil and Water Conservation* 42(3):144-149.

The Leopold Legacy for Soil Conservation

Melville H. Cohee

Two great leaders in conservation met in 1933—Hugh Hammond Bennett and Professor Aldo Leopold. As chief of the new Soil Erosion Service (SES), Bennett was seeking advice on watershed demonstration projects for soil erosion and flood control. He had appealed to most state land grant colleges and universities. Officials at the University of Wisconsin responded early on. Leopold was part of the small delegation from the university that traveled to Washington, DC, to counsel with Bennett.

Previously, Leopold had favorably impressed R.H. Davis, the new SES regional director in Lacrosse. Leopold had convinced Davis that wildlife management should be an integral part of the national SES program and each watershed demonstration project. Bennett likewise readily accepted this advice because he wanted all aspects of good land use to be included in the demonstrations. He believed that each acre on a farm or ranch should be used for and treated in accordance with its capabilities. Bennett loved the land and understood its many features; Leopold held deeper ecological-based perceptions about land use.

I believe that Aldo Leopold saw a sizable opportunity and hope in Bennett's comprehensive farm conservation plan approach through the demonstration projects (and later through conservation districts) for extending the concept of land husbandry. In fact, he discussed this during his many seminars with us young SES staff people at Coon Valley, Wisconsin. He wanted to see developments go beyond "conservation treatments as superficial" and become effective realizations that "in land, just as in the human body, the symptoms lie in one organ and cause in another." He did not want soil conservation practices that, to a large extent, would be only local alleviations of biotic pain as he called it. He knew the difference between "land doctoring" and the "science of land health." His port of entry was through the integration of biology (via wildlife management) in plans for farms and other watershed land.

Leopold and Bennett were indeed co-fathers of the wildlife management aspects of modern-day soil conservation on farmland. It was Leopold who integrated wildlife management into the nation's first watershed demonstration project in Wisconsin's Coon Creek watershed.

Leopold also spent many days and nights in the Coon Valley project area in 1934 and 1935. In the beginning he engineered most of the wildlife management work for the project. The first biology-wildlife manager on the Coon Valley staff was knowledgeable but inexperienced. Habitats for wild animals needed study and inventory. With such information each comprehensive farm conservation plan could be developed to encompass wildlife aspects. It was Leopold who outlined how all of this should be done.

As the project economist responsible for balanced coverage in the farm conservation plans, I had to learn as much as possible about wildlife management principles. In so doing I took part in many survey operations. Moreover, in staging farmer meetings about the project and its operation we went beyond physical and economic considerations to get across some of the esthetic values to be gained—what scholars today might include as ecological and ethical considerations.

Most off these insights were gained from Leopold, especially in the many "talk sessions" our small staff had with him after supper and well into the night. Determinations were high to make all aspects of the farm conservation plan successful in operation. This was true for the wildlife management parts, like all others. When deep snows came in the first two winters, food stations were established to help prevent undue mortality among wildlife. Again, it was Leopold who engineered this emergency step, which later was not needed to supplement in-field food and cover left by farmers.

The concepts applied in the Coon Creek project spread into many additional projects throughout the United States and continued in the early work of conservation districts. But then the principles and determinations of the Coon Valley days began to wither. Pressures were great for large numbers of farmers to participate in what could loosely be characterized as soil conservation; only parts of a farm were planned and selected practices applied. One of the first exclusions was the exacting attention to Leopold's biological entry, through the wildlife management door.

Fifteen years after his first Coon Valley implants, Leopold in *A Sand County Almanac* offered a final, perceptive plea for conservation: "Conservation is getting nowhere because it is incompatible with our Abrahamic concept of land. We abuse land because we regard it as a commodity belonging to us. When we see land as a community to which we belong, we may begin to use it with love and respect. There is no other way for land to survive the impact of mechanized man, nor for us to reap from it the esthetic harvest it is capable, under science, of contributing to culture."

The 1985 farm bill may correct some of this slippage. But unless implemented with Leopold's principles and the Coon Valley experiences in mind, little will be gained beyond the notion of "land doctoring." It is not easy to apply in practice those teachings from Leopold's "science of land health."

Everyone who knew Aldo Leopold had a feeling of kinship with a master. His conversations effused sincerity and love for nature and intolerance for undue harm or destruction of natural resources. Like many great leaders, it is often lamented that "the mold was lost when he was born," 100 years ago. Perhaps a successor will arise and assume leadership in continuation of the land ethic trail he blazed.

ORIGINAL PUBLICATION DETAILS

By Melville H. Cohee, USDA Soil Conservation Service (retired), Oregon, Wisconsin.

Cohee, Melville, H. 1987. The Leopold legacy for soil conservation. *Journal of Soil and Water Conservation* 42(3):142.

Part 3
Development of Conservation Thought
on Land Stewardship
and Natural Resource Values

The Origins of Conservation Philosophy

Clarence J. Glacken

Historically, the view of the earth as a planet designed and made habitable for all forms of life has involved two attitudes toward living nature: It is beautiful and it is useful. In Christian thought there has been nothing accidental in the combining of these attributes, for the earth was beautiful because it was an expression of the perfection of the creator, and it was useful to man because he could get from it physical sustenance and the religious inspiration to worship his creator. This conception of the earth was a dominant one through the Middle Ages and early modern times and was still important in the nineteenth century, as shown by the number of natural theologies written by scientific men, such as the geologists John MacCulloch (1837) and Hugh Miller (1857), although it had come under increasing criticism beginning in the eighteenth century because of the teleology which was assumed in the argument from design. As a consequence of this belief, the Western idea of man as part of nature has never been quite forgotten despite opposing tendencies such as the divorcement of humanistic and scientific inquiry and specialization within disciplines.

There was one significant assumption in this religious conception of nature: In the natural order of things, men would use their environment, changing it for their own ends and improving on its natural state; in descriptions of nature which appear in works with this theme, both pristine and domesticated nature are described, the descriptions often intermingling: The forests of the mountains are set off from the tilled fields of the lowlands; the olive or the fruit trees form neat rows on the borders of wild-growing low-lying shrubs; the peasant huts are built along the stream; the domestic animals tended by their herders graze in the forest.

In the writings from the seventeenth to the nineteenth century which were devoted to the design argument, whether in natural theology, philosophy, natural history, geography, biology, or other disciplines, nature was conceived of as a usufruct and man as the highest being in the creation had responsibilities as well as privileges in using it. His role with regard to the rest of creation was often looked upon as that of a steward or a caretaker of God. In the works of such physico-theologists as John Ray and William Derham, man has the role of a partner of God in improving primeval nature, in conversions of natural landscapes for his own purposes, in agriculture, in the clearing of forests, in the drainage of swamps: His improvements in soil fertility through the use of manuring and marling were other instances of the partnership of man and his creator not only in maintaining but in improving the earth, through his inventiveness, for his enjoyment and use and for the greater glory of God (Ray 1759; Derham 1798).

In addition to these ideas, there were others of a more humble and practical origin which in their cumulative effects had a great influence on the development of conservation, ideas based on observations that human activities brought about undesirable changes in nature. These observations accumulated not in philosophies and in natural theologies but in learned and technical works; one sees them in increasing number in the eighteenth century. It is these two trends, one, aesthetic, philosophical and religious, the other, practical and technical, that have characterized ideas of conservation throughout its history.

Forests Were Recognized in Late Middle Ages

These notices, mostly descriptive of local conditions, began to accumulate in the late Middle Ages. For the most part they concerned the use of the forests. Concern about them was not a new phenomenon; one has glimpses of it in classical and post-classical times, in the legislation of Roman, Byzantine, Merovingian, and Carolingian times. The forests of Europe were related intimately both to the economics and the amenities of life, and there were conflicts of interest in using them. The most obvious—and the one that has been emphasized in modern times—was the indirect conservation of the forests because they were royal or noble hunting reserves. They were also the source of wood fuel to keep kilns and smelters going, and of wood for building purposes. Forests were often centers of the beekeeping industry which was important before sugar was known, and there was a lively trade in beeswax for making candles used by monasteries in the forest wilderness. Forests were grazing lands and the early attitudes toward them, and the legislation enacted regarding them, cannot be understood without realizing that their importance as grazing lands for sheep, cattle, goats, and especially for swine often was equal to or greater than their significance as wood producers. It was the complex role of the forest in the economy which brought about laws and regulations governing the uses of forests in every European country (Maury 1856; Schwappach 1886-1888).

The practice of transhumance often meant clearance of forest lands to increase mountain pastures at the expense of trees. There were conflicts too between farmer and herder in the width and use of sheep runs through agricultural land. It was, however, the extensions of agriculture and industry at the expense of the forests that were most conspicuous: Agriculture meant cutting and burning of trees, the assarts, the Brandwirtschaft that continued in some parts of northwestern Europe, in Finland for example, until the latter part of the nineteenth century. Furthermore, wood and water were the great sources of energy before the age of coal. Another explanation for the growth of the conservation idea thus is to be sought in conflicts of interest with regard to use. There are two famous illustrations of these conflicts, John Evelyn's *Silva* which was first published in 1664, and the celebrated forest ordinance of Colbert of 1669.

In his *Silva*, a landmark in the history of silviculture, Evelyn describes the encroachments of agriculture, the dangers of grazing and of metallurgy; the competition in land use can plainly be seen. Agriculture, Evelyn said, was extending itself at the expense of the forests, and the iron mills were destroying them because of their demands for charcoal to reduce the ore and in the working of iron. Evelyn's remarks reveal the existence of problems posed by the innovator and the conserver: The distribution of minerals which a culture finds useful, even indispensable, conflicts with the preservation of the forest whose existence is governed by entirely different physical circumstances. Early modern conservation ideas are really a recognition of the fact that the distribution of mineral deposits with relation to forests, even with the industrial development existing before the Industrial Revolution, meant the slow, relentless destruction of nature (Evelyn 1786).

Resource Decrees Date to Thirteenth Century

One of the merits in studying the history of legislation is that laws are excellent indications of the abuses which provoked their enactment; they are less trustworthy guides regarding accomplishments because of the gap between intent and enforcement. Colbert's famous declaration that France would perish for lack of wood was not a sudden outcry against forest destruction; it was a condemnation of practices which had been commonplace in France for centuries, for the ordinance, approved by Louis XIV, was the latest and most complete of a series of decrees dating back to the thirteenth century. One general provision of this long law sums up rather well the problems it was trying to solve: "We forbid them [the Grand Masters of the Forests] to permit or suffer any kilns, furnaces, charcoal-making, grubbing and uprooting, lifting or removal of beacons, acorns, and other produce from our forests, contrary to the provisions of these presents, on pain of arbitrary fine, and reparation of all our damage and interests" (Brown 1883).

From the late eighteenth to the mid-nineteenth century, the literature relating to conservation increased both in scope and in depth; there has been no break in the continuity since those times. French, Italian, German, Austrian foresters, engineers, and hydrologists studied the problem of deforestation as the cause of Alpine torrents, among whom the names of Fabre and Surell are perhaps the best known. In this period the long discussions concerning deforestation and climatic change were pursued with vigor, including among the scientists such famous names as von Humboldt, Boussingault, and Becquerel (Marsh 1864; Brown 1880; Becquerel 1871). New works on land reclamation and the making of polders—the Dutch with their long experience had already created an extensive literature—were part of an effort to extend these techniques throughout Europe to reclaim land. Studies on the fixation of sand dunes, important because the encroaching sands inundated villages, were made from the dune areas of the Atlantic and North Sea shores to those of Pomerania and East Prussia. Fixation through the planting of maritime pines had already been undertaken in the Landes region of France in 1795. Scientists began investigating soil erosion of the Mediterranean basin and of Asia Minor; Carl Fraas, for example, made an extensive study of the deterioration of the lands of Asia Minor and the eastern Mediterranean, ascribing it to the loss of vegetative cover (Fraas 1847).

First Reference Work on History of Conservation

Despite its volume and scope, there was no real synthesis of this literature until

George P. Marsh published in 1864 *Man and Nature; or Physical Geography as Modified by Human Action.* Marsh had an impressive command of his subject, the result of travel, observation, and exhaustive reading of the contemporary literature; he documented his work in modern fashion, and, in addition to its other merits, it is a valuable reference work on the history of conservation. Marsh's main examples were the countries of northwest Europe and the Mediterranean. In making a study of lands which had a long history of human settlement, Marsh warned his country-men—by comparing environments of old with those of newly settled regions—of the dangers of disturbing the balance of nature. The lands of the New World would require careful husbanding if they were to remain habitable for future generations. He wrote his work to show the immense power exerted by man throughout his-tory in modifying the natural environment, discussing his role in plant and animal domestication and in the activities of domesticated animals like sheep, goats, and camels, really extensions of human activity; the consequences of deforestation, Marsh's chief concern, for it was the preoccupation of many scientists of that period; man's effects upon the waters and upon the sands.

The importance of Marsh's work from a conservation point of view is that it clearly demonstrated that conservation was an historical, cultural, and philo-sophical problem: throughout history the earth had been changed by peoples who neglected or ignored the effects of their destruction of the natural environment. It is noteworthy that virtually all of his examples refer to the period prior to the Industrial Revolution.

In the work of Marsh, the idea of an organized and harmonious nature designed by the Creator for the use of all organic life and with man at the apex of the scale is merged with this vast technical literature: From it came a philosophy of nature and of man's responsibility in maintaining it as a usufruct. Man should be faithful to his stewardship, for the earth was beautiful and useful because the Creator had intended it to be so.

Long before Marsh's time, however, American agricultural writers, naturalists, and farmers had already observed and written about deforestation, soil exhaustion, and soil erosion. Some of this literature was inspired by English writings, but much of it came from the observation of local conditions. Among the most well-known of these early conservationists were Jared Eliot, Thomas Jefferson, John Taylor, and Edmund Ruffin (McDonald 1941; Bennett 1944; Chinard 1945).

Another American, Nathaniel S. Shaler, was an outstanding figure in the last quarter of the nineteenth century because, like Marsh, he looked at conservation from a world point of view and because he was interested in so many phases of the subject. Shaler, a geologist, expressed a conservationist philosophy similar to that of Marsh, but he emphasized more than Marsh had the importance of understand-ing soils and the seriousness of soil erosion. In an article which appeared in the *National Geographic Magazine*, Shaler, who was no pessimist, warned his coun-trymen in vigorous language of the danger that their soils would wash away if they neglected to care for them. His monograph on soils was prepared to assist anyone desiring an understanding of soils, an essential study he thought for settlers in a new country whose traditions concerning the soil of a district were comparatively meager. Shaler was a man of very broad interests; in his later writings, he presented one of the earliest and most interesting analyses of mineral consumption by modern civilization, and he advanced ideas concerning the resources of the seas, future energy sources, and the changes which man, with his growing numbers would continue to make on inorganic matter and organic life on earth. Shaler represents a shift from the old emphasis on deforestation, characteristic of Marsh and his predecessors, to a new one, concerned with the contemporary and future uses of the forests, soils, and minerals (Shaler 1891, 1896, 1906).

Exploitation of Land Resources

The expanding economies of the latter part of the nineteenth century produced a correspondingly expanding literature on the need for conservation. A German economic geographer, Ernst Friedrich, popularized the word Raubwirtschaft (plunder economy), attempting at the same time to find a rationale in man's exploitation of the world and finding it in the formula that unwise destructive exploitation led to poverty and distress which in turn would lead to rational exploitation (Friedrich 1904). The British parliament investigated the causes of deforestation in India. Studies of the tropical plantation, of shifting agriculture and grazing in the tropics were beginning to appear. There was a growing literature of protest against the wholesale killing of the plume birds, the bison, the big game of Africa, seals, and whales; one writer referred to the times as an "age of extermination."

In the United States, the rapid extension of settlement had brought about an awareness, among men interested in conservation, of the changes which were transforming the country. It was different from old Europe where the contrast between virgin land and settled land was deep in the past. In America the contrast was real to a single generation: In the sixties Marsh was warning of possibilities which were realities in the nineties. There is a vigor, a comprehensiveness, a sense of urgency in this late nineteenth and early twentieth century American literature: the tone is unmistakable in the early report of Frederick Hough, in the report of the National Conservation Commission which followed the White House Conference of 1908, in the proceedings of the first Conservation Congress in Seattle, and in the pages of Van Hise's famous textbook (Hough 1878; National Conservation Commission 1909; National Conservation Congress 1910; Van Hise 1910). There is an emphasis on the need of maintaining the balance of nature, on developing resources without waste, on the intimate association between a nation and its resources (perhaps nowhere more eloquently expressed than in this literature) and the need of safeguarding and of restoring the natural beauties of the country. Although these investigations had their origin, as Gifford Pinchot (1937) has said, in the movement for forest conservation which is so closely associated with his name, it should be remembered that forestry was a broad term. These men were interested in the forests, in the streams, in the grazing lands, and in the soils; the idea of conservation had expanded to include the material resources and values. There was a great deal of optimism in America during this period, but a few men also took advantage of the unparalleled opportunity to soberly study the immense power which modern societies had acquired to change their physical environments.

Summary of Contemporary Conservation Ideas

Although it would be rash to attempt in this brief space a summary of contemporary conservation ideas, a few general observations may be helpful in understanding the literature of today and how it differs from that of the nineteenth and the early part of the twentieth century.

In the first place, much more is known about the nature and distribution of the world's soils. Speaking generally, from classical antiquity to the time of Liebig in the middle of the nineteenth century, the main preoccupation of students of soils was with the maintenance of the fertility of arable lands; soil science, closely identified with agricultural chemistry, was practical in its goals and in its research. In the late nineteenth century, the work of Hilgard in America and of Dokuchaiev and his school in Russia called attention to the need of studying soils in their own right and independently of the practical considerations of agriculture. From these investigations of little-worked or virgin soils of the United States and Russia our modern concepts of the nature of soils have been derived, and soil science today is much more closely related to conservation than it was in the last century.

In the second place, the ecological point of view is much more in evidence today than it was in the past. The idea of a close and interlocking relationship in nature is an old one; its origin can be traced to the idea that nature was a divinely arranged harmony. The concept of the unity of nature, however, was given more specific meaning by von Humboldt's studies of plant geography, by Darwin's concept of the web of life, and by many of the late nineteenth century ecologists. Modern studies of the populations of biotic communities, of the history of vegetational change, and of the effects of cultural changes in biotic communities have done much to broaden the theoretical bases of conservation, even though there has been criticism of some ecological theory such as the succession and the climax concepts.

In the third place, the American dust storms of the thirties have had a great influence on conservation ideas because they occurred in combination with the depression and the widespread migration from the Dust Bowl. Their effect was to revive an old but neglected point of view that conservation was deeply involved in history and in the cultural values of a society. The surveys of Bennett and of Jacks and Whyte which followed called attention to the world-wide nature of these problems (Bennett 1939; Jacks and Whyte 1939).

Since the end of World War II, a new combination of old ideas has resulted in a greater emphasis on the power of human cultures in changing the environment, on the relation of custom and law to conservation, and on the increased growth of the world's population. There have also been renewed attempts to create a conservation ethic and philosophy, a moral and aesthetic attitude toward nature which, I cannot help but feel, is the modern substitute for the earlier scientific and religious belief that the world of nature with all its beauty was a harmonious whole designed by the Creator.

References

Becquerel, Antoine César. 1871. Forests and their climatic influence. Annual Report of the Smithsonian Institution for the Year 1869, 394-416. Washington, DC.

Bennett, Hugh Hammond. 1939. Soil Conservation. New York and London.

Bennett, Hugh Hammond. 1944. Thomas Jefferson, Soil Conservationist. USDA Miscellaneous Publication 548. Washington, DC: USDA.

Brown, John C. 1880. Reboisment in France. London.

Brown, John C., trans. 1883. French Forest Ordinance of 1669; with Historical Sketch of Previous Treatment of Forests in France. Edinburgh.

Chinard, Gilbert. 1945. The American philosophical society and the early history of forestry in America. Proceedings of the American Philosophical Society 89:444-488.

Derham, William. 1798. Physico-Theology: or, A Demonstration of the Being and Attributes of God, from His Works of Creation. Vol. II, 143-165. London.

Evelyn, John. 1786. Silva; or, A Discourse of Forest Trees, and the Propagation of Timber in His Majesty's Dominions. Vol. II, 256-269.

Fraas, C. 1847. Klima und Pflanzewelt in der Zeit, ein Beitrag zur Geschichte beider. Landshut.

Friedrich, Ernst. 1904. Wesen und geographische verbreitung der "Raubwirtschaft." Petermanns Mitteilungen 50:68-79, 92-95.

Hough, Franklin B. 1878. Report Upon Forestry. Prepared under the direction of the Commissioner of Agriculture, in Pursuance of an Act of Congress Approved August 15, 1876. Washington, DC.

Jacks, G. V., and R. O. Whyte. 1939. Vanishing Lands: A World Survey of Soil Erosion. New York.

MacCulloch, John. 1837. Proofs and Illustrations of the Attributes of God, from the Facts and Laws of the Physical Universe: Being the Foundation of Natural and Revealed Religion. London.

Marsh, George P. 1864. Man and Nature; or, Physical Geography as Modified by Human Action. New York.

Maury, Alfred. 1856. Les Foréts de la France dans l'Antiquité et au Moyen Age; Nouveaux Essais sur Leur Topographie, Leur Histoire et la Législation Qui les Régissait. Paris.

McDonald, Angus. 1941. Early American Soil Conservationists. USDA Miscellaneous Publication 449. Washington, DC: USDA.

Miller, Hugh. 1857. The Testimony of the Rocks; or, Geology in Its Bearings on the Two Theologies, Natural and Revealed. Edinburgh.

National Conservation Commission. 1909. Report of the National Conservation Commission. Senate Document 676, 60th Congress, 2d Session.

National Conservation Congress. 1910. Addresses and Proceedings of the First National Conservation Congress Held at Seattle, Washington, August 26-28, 1909.

Pinchot, Gifford. 1937. How conservation began in the United States. Agricultural History 11:255-265.

Ray, John. 1759. The Wisdom of God Manifested in the Works of Creation, 12th ed. London.

Schwappach, Adam. 1886-1888. Handbuch der Forst und Jadgeschichte Deutschlands. Vol. I, 26-53, 156-178.

Shaler, Nathaniel S. 1891. The origin and nature of soils. Twelfth Annual Report of the United States Geological Survey, 1890-91. Part I–Geology, 213-345. Washington.

Shaler, Nathaniel S. 1896. The economic aspects of soil erosion. National Geographic Magazine 7:328-338, 368-377.

Shaler, Nathaniel S. 1906. Man and the Earth. New York.

Van Hise, Charles R. 1910. The Conservation of Natural Resources in the United States. New York.

ORIGINAL PUBLICATION DETAILS

By Clarence J. Glacken, Assistant Professor, Department of Geography, University of California, Berkeley, California.

Glacken, Clarence J. 1956. The origins of the conservation philosophy. *Journal of Soil and Water Conservation* 11(2):63-66.

Conservation: A Social and Moral Problem

Carl C. Taylor

"**M**ankind survived tens of thousands of years without facing a series of conservation problems," said Dr. Julian Stewart, head of the Department of Anthropology, Columbia University, at the Inter-American Soil Conservation Conference at Denver in 1948. He added that "until the European Conquest, land use tended to conserve rather than destroy resources." In this article it is my purpose to attempt to do a very simple yet a very difficult thing, viz.: first, describe why and how those countless generations of men to whom Dr. Stewart referred conserved their natural resources; second, analyze the moral imperatives which controlled their behavior; and third, say something about the issues and problems of developing moral imperatives which will today and in the future cause us to conserve our natural resources.

Plato asserted that "no group, even the most antisocial, can exist without a minimum of morality necessary for its survival." Darwin, a biologist who spent his life studying lower organisms than man said he discovered mutual aid everywhere in the struggle for existence. He imputed this to what he called "social instincts." Morals, however, are not instincts which can be biologically inherited; they are human sentiments which must be individually learned and socially enforced. Morals (the mores) are developed by all groups and societies to sustain what each society believes is right and good. The most stern morals of all groups and societies have been, and are, those which safeguard and promote the survival of the group.

If we and other peoples are today destroying the resources upon which our survival depends, it is either because we do not know that our survival depends on them or because we have not developed social and individual sentiments which have risen to the level of moral imperatives. Simple, so called "primitive," groups did know that the resources under their control were limited and that they had to conserve them or die. They therefore developed deep moral convictions about conservation and passed these mores on to each succeeding generation. Dr. Stewart said "disregard for the future is a comparatively recent phenomenon." We seem to have developed this disregard in a big way, but I am convinced that it is not because we are inherently less moral than primitive man. It is because our human relations are so complex that we don't really know all we should about them. Morals are the cement of group life and we have difficulty in creating this cement because we aren't intelligently conscious of all the groups to which we belong. We can't develop moral imperatives about our group problems until we develop social sentiments within these groups and we won't develop these sentiments until we become aware that we are members of groups which are far greater than those in which we participate in a face-to-face fashion.

Once simple societies had established their moral imperatives they enforced them by sanctions and taboos. They didn't have to depend much on laws because each individual believed that to violate a taboo was an unpardonable sin. Rivers says, "among such people as the Melanesians there is a group sentiment which makes unnecessary any social machinery for exercising authority." Another anthropologist tells of seeing a Copper Eskimo, who because he was ill, felt sure he had unknowingly violated a taboo. He went to the Medicine Man who worked over him for two days to help him get right with his own conscience.

The Iroquois Indians depended heavily on fish as a part of their food supply and they guarded this natural resource by a taboo against fishing during the spawning season. The individual Iroquois believed if he violated this taboo that Orenda, the Spirit of the Great Wacunda, would strike him dead or bring some dire thing into his life. We call such beliefs superstitious, and so they were, but the point is that they worked because they were group moral imperatives and as such were a part of the conscience of each individual member of the group.

Taboos and sanctions were supposed to identify those objects and those types of behavior which were good or bad. The most universal ones had to do with the conservation of the natural resources upon which the physical survival of groups depended. Others sought to perpetuate the integrity, or wholeness, of the individual groups. Their methods of enforcing their moral imperatives were cruel but they nevertheless converted their beliefs into religious doctrines and their practices into religious rites the better and surer to preserve them. In order to preserve the known natural resources upon which their survival depended, they practiced infanticide. They allowed their aged to die or be destroyed when they could no longer contribute to group survival. In the most dire situations, they ate their own parents and believed that neither their parents nor they could go into the Great Beyond unless they did eat them. Crude as such things were, they had to do with the necessities of group survival and therefore became part of the moral imperatives of their groups, which no one violated. As the organization of simple societies rose above the level of primitive communism, in which there were no specialized functions and few if any specialized obligations, the first step in the direction of what now prevails universally in modern society was the gradual appearance of specialists. The first specialists to arise were the religious leaders, medicine men, and priests. They had two basic functions; one was to be the keepers and the teachers of the mores and the other was to be liaison between their people and the Spirits. As keepers and teachers of the mores, they passed the moral imperatives of their group down from generation to generation. Acting as liaison between the people and the Spirits, they interpreted to their people what the Spirits desired, and they were believed to be on good enough terms with the Spirits to influence them on behalf of their people.

Again we say such beliefs are superstitious. Sure they are but they worked and all I am saying is that by discovering on what basic issues of life people over long periods of time have developed their moral imperatives and by learning how deeply they have been convinced that both group survival and group integrity depended on the maintenance of these moral imperatives, will help us more clearly to see our own situations and responsibilities. We are not going to turn back the direction evolution has traveled from the ways of primitive to the ways of our secular society. But it is my belief that some guideposts for underwriting our secular behavior with moral imperatives can be visioned by finding answers to the question of how the sure moral imperatives, which at one time existed, were diluted during that evolution. I believe the first step in that dilution came when social contacts were extended beyond the boundaries of face-to-face groups and thus included strangers. A stranger was not a real person because he was not a member of one's own group. The second step was the development of money economy, ultimately applied to natural resources as well as trade. When land was alienated from group ownership and became an individually merchantable good, the value of land was no longer measured by group survival values but by money values. The third step was the development of specialization, i.e., the individualization of personal functioning, in which the value of each person was measured in terms of his individual talents and not in terms of good group membership. All of these were steps in the long evolution of the secularization of human relations and the consequent dilution of the individual's group responsibility.

New specialists in trading, a merchant class, early came into existence whose members did nor live directly from the land. Farm products to them were not the fruits of natural resources; they were merchandise. Under the impact of a developing commercial economy, ownership of land and other natural resources was alienated from group ownership and came to be exploited as private enterprises. The Aztecs went only part way in this direction. They allotted lands to individuals but took them away at the end of two years if the lands were not properly handled. In deep peasantry the land is individually owned but it is "not for sale" because it must, by their mores, be passed on to the next generations. As a family inheritance, it isn't exploited, but conserved. The greatest deterioration of natural resources has not occurred and is not occurring where traditions of this type prevail. It isn't occurring where groups of people must still live on limited resources. It is occurring where individual entrepreneurs don't recognize the life and death struggle between group survival and resource depletion.

A number of peoples are today in the process of these transitions. Guatemalan Indians who lived for hundreds of years in their self-sufficient villages are entering the marketing economy. Soil erosion is probably traveling fifty times faster because they are pushing production for market sales. The same thing happened all over the world. It was not primitive men who destroyed the ancient civilizations but men and societies who came on the scene after primitive societies with their stern conservation mores and before modern men and societies had developed a conservation conscience.

Our own history of handling our natural resources is the best laboratory in the world to study and discover what happens when the transitions which I have been discussing take place rapidly. It was not only by design but by fortuitous good fortune that in the settlement of this continent our forefathers broke almost completely with the traditions and social controls of older civilizations. They were escaping from the tyrannies of such controls and they fell heir to the most giant body of undeveloped natural resources in the world. It was an historic accident that the settlement of our lands and the exploitation of our natural resources was coincident with the development of the Industrial Revolution. Trade and commerce opened the markets of the world for the products of our natural resources and the opportunities of our vast frontiers invited men with initiative and enterprise to exploit these resources.

In this miraculous development we demonstrated what we thought were the successful ways of life and naturally built creeds to buttress and promote that way of life. It is easy to understand why we built and believed in these creeds. For three hundred years we developed by expansion of our frontiers. We discovered and exploited unheard of natural resources. We found these resources by trial, error, and success methods; took chances, knowing that if a man missed the first time and at the first place there were plenty of opportunities to try again. Some speculation was necessary under the circumstances and for 300 years such speculation paid off in a big way. Thus Expansion, Exploitation, and Speculation almost automatically became our creeds.

Out of these experiences and developments we built the strongest and freest nation on earth, but today we are raising questions. We are asking whether the experiences of building America were not so abnormal (super-normal) that they cannot be the pattern for all places and peoples of the earth, or even for our own future. We are asking whether our creeds of expansion, exploitation, and speculation will work as well in our maturity as they did while we were in the process of growing up. There is some evidence that we are even beginning to answer these questions by saying that we are now trying to develop both the technique and creeds of conservation and security. If this be true then we can utilize the lessons from history of how peoples of the past used moral sentiments to buttress and promote their beliefs and practices.

The development of moral sentiments is more difficult in a highly complex and impersonal society than in simple face-to-face groups, but the same basic principles apply to both. As simply stated as possible the principles are: (1) All groups must have morals (mores) in order to survive as groups. (2) The morals are automatically maintained and sustained when and where the personal consciences of the members of the groups are the same as the mores of the group. (3) Penalties are assessed against, or other group controls are exercised over, those members of the group who violate the group's morals.

But the application of these principles is difficult in a big complex impersonal society. Human relations are so atomized that they become secular, not sacred as they are in family life, and as they were in whole groups of simple societies. Today we are members of groups that are so large that we don't understand, or even know, all the relationships in which we are involved. We are members of a great many groups, some of them groups within groups, and our group loyalties sometimes conflict with the others. Primitive groups were not involved in these difficulties. Every one knew everyone else, knew in what ways his behavior affected all others and the ways others' behavior affected him. Because of this intimacy it was not too difficult to develop moral sentiments about all these relations and not too difficult for the group to enforce the moral codes which these sentiments dictated.

Moral sentiments in all history have been the product of group necessity and the acceptance of group responsibilities. It cannot and will not be different in the future. The conservation of our natural resources will not be recognized as a moral problem by any other process. Those of us who are students of natural resources and those of us who are students of human relations must join hands in making the necessities of conservation widely known. Religious leaders, scientists, and statesmen must join hands in making individuals, groups, and governments willing to assume responsibilities for promoting and, if necessary, enforcing responsibilities. No matter how many or how complex the human relations men now have, they must understand them and be willing to share the responsibilities which they involve. Only thus can they be intelligently moral because only thus can they know how to be moral. The need is not only for scientific knowledge but conscience, not only for reason but for sentiment. Only when men by the millions become rationally conscientious about the basic issues of human survival will such giant problems as conservation be solved by moral principles.

If the problem of survival is even in part the problem of conservation, then conservation is a moral problem. The issue is how can we make men by the millions conscious of and concerned about this fact. We can't do it by passing resolutions. Moral sentiments in all history have been the product of recognized necessity and the acceptance of group responsibility. It cannot and will not be different with conservation. The task no matter how difficult dare not violate the principles of which all through this article I have been giving illustrations. Morals don't arise in a complex society such as ours by superstition and are not enforced by ignorant medicine men. They arise by such processes as described in the *Journal of Soil and Water Conservation*. The morals of one age are not necessarily the morals of all succeeding generations but their functions are the same in all societies. Modern scientific men don't eat their ancestors as moral imperatives but they must do what their reasons and consciences dictate. They must be rationally conscientious men, and there must be millions of them who are willing to measure what they do by what they ought to do.

ORIGINAL PUBLICATION DETAILS

Carl C. Taylor, Head of the Division of Farm Population and Rural Life, Bureau of Agricultural Economics, Washington, DC.

Taylor, Carl. C. 1951. Conservation: A social and moral problem. *Journal of Soil and Water Conservation* 6(1):7-9,14.

Relationship of Conservation Attitude to Science and Private Enterprise

Paul B. Sears

Editor's note. This piece—representing a distinctly post–World War II view of conservation and its relationship to science, technology, and private enterprise—was initially presented at the Annual Meeting of the Soil Conservation Society of America in St. Louis, Missouri, November 1949.

It is important to remind ourselves that much of the best conservation is older than science, and that even today technical knowledge is seldom the limiting factor in our wise use of natural resources. I recently saw the collection of aerial photographs taken by the 37th Division for its advance from Manilla to Baguio. These photographs show amazingly perfect terraces developed on steep hillsides, with minimum injury to the surrounding tropical forests. This was the work of so-called primitive people. And in Mexico I have examined small remnants of good land use practice which long antedate the Conquest. These remnants may fairly be called living fossils of sound cultural adjustment which have somehow withstood the impact of modern exploitation.

I would by no means underestimate the very real progress of the past twenty years in the United States. Public opinion has been aroused, and public support organized to a degree that would have seemed fantastic and hopeless in 1930. Sportsmen, for example, who have to put up much of the cold cash for conservation, have been literally reeducated and are now seeing their particular problem as merely an aspect of wise land use. Finance and industry are stepping into the picture and lending substantial support not only to the conservation of their own raw materials, but to more general enterprises. School administrators are waking up and the next yearbook of their national organization will deal with conservation education. I could continue for some time with such optimistic reports.

Is there any reason, therefore, why we should not use our time here in mutual self-congratulation and return to our tasks refreshed for simply carrying on as we have begun, with more vigor than ever? I wish it were possible, but I doubt it. For it seems to me that, in any fundamental sense, we have barely scratched the surface.

I mention the good conservation of prescientific days not to discount the importance of science, but to emphasize the fact that conservation is basically a moral and social problem. Here science has two really great contributions to make. First of all it must give mankind a realistic notion of the kind of universe in which we live. Throughout human history conduct and the values which regulate it shape themselves according to the kind of world in which men believe themselves to be. The second great role of science is to give us the technical means to carry out the implications of our fundamental beliefs.

A much less important function of science (desirable though it may be) is to lift the physical burdens from the shoulders of humanity and to provide comfort and diversion. And finally the prostitution of science is to employ it in destruction, by marshalling natural resources for wars that can by any possibility be averted.

If I am correct in these assumptions, certain conclusions follow as a matter of course. The scientific interpretation of the universe pictures it as the expression of universal law, which plays no favorites. Conservation of energy and matter are at the very basis of this great system. The science of ecology sees living communities and the soil which sustains them as capable of building up to a thermodynamic steady state, whose organization permits the maximum efficiency in the use of solar energy and inorganic materials to support life. And it sees that man, on a wide scale, has interrupted the course of this development without substituting its equivalent.

This being so, it is fair to ask: Do we act as though we believed it? Does our scheme of values really impose any obligation, any emotional sanction, to act as though this were true? If the sense of obligation does exist, it can be phrased very simply. The Pennsylvania Dutch farmer, who never heard of thermodynamics, expresses the principle very plainly: "I want the boy to get this farm as good as I got it." With him this is a matter of principle, of ingrained obligation.

We cannot go back to the simplicity of the Chinese peasant, the ancient Peruvian terrace maker, or the old Amishman. Our problem is far more complex. Ultimately it is a problem of artistic insight and spiritual leadership. But I think we can lead most promptly to these goals by making the American people scientifically literate.

Let me be very explicit on this point. A man can be in the very forefront of a scientific field and still not be scientifically literate. I could name one such individual, of great influence but utterly ignorant of biological processes. He actually believes that all conservation talk is nonsense, and that whenever the situation get really serious he and his kind will come up with a new solution.

To counter such misbelief, I think we must learn that true science instruction must begin with the very earliest education. Words must be learned, not in vacuum, but in relation to the objects and processes which they symbolize. And our science teaching at higher levels must be drastically overhauled. Survey courses were an attempt in the right direction, but they failed because they left out or slighted the physical experiences with the world of nature which are essential to scientific literacy. We call this experience laboratory work, but the average elementary college laboratory is at present hardly a place for valid experience, let alone adventure and discovery.

Most college instruction in science is frankly shaped to the imagined needs of the future specialist—less than ten percent of the whole student group. Once we can send out future professional men, business and industrial leaders, poets, artists and writers, who sense in their very bones the universal character of scientific law, I believe they will quickly enough begin to shape our culture to fit that conception.

Granting this as possible, there remain tremendous obstacles in the way. Our civilization as it is—even in this world revolution this is going on—has tremendous inertia and will be hard to modify. We have been applying science most powerfully to two ends that are certainly not conservative. One is the elaboration of consumers' goods, with attendant marketing pressure and without much reference to raw materials. The other is the related problem of making war munitions. These applications are out of all proportion to other uses of science, even medical science. You may have noticed that Billy Rose was shocked at the salaries available for research workers on poliomyelitis and cancer. The distributary flow of money is not a complete measure of the aims of a culture, but it is pretty significant.

The era of ugly architecture in the United States coincided with the rape of the white pine forests of Michigan and Wisconsin. This was accomplished by a tremendous sales promotion campaign in which free plans were given out showing how to use wooden gingerbread in the most extravagant manner possible and appealing to customer vanity in thoroughly modern fashion.

That sales promotion has done much to raise the standard of living I do not question. But the mere possession of consumers' goods is a matter of rapidly diminishing returns and is no guarantee of human happiness. I do not envy the situation of big industry whose employees strike for higher wages partly because they can read the clever advertisements of big industry. I know that extravagant use of such things which are not necessary is pretty tough on natural resources.

Nor do I have any quick remedy for this unbalance. I should, however, like to see thrift encouraged by something more inviting than a bleak box ad of the local savings bank, "1% interest on time deposits." I should like to see, too, more statesmanship on the part of leaders in labor, finance, and industry—more concern with general welfare than is now visible to the naked eye. And I question whether we shall ever have anything like a stable economic system until men who work at machines can feel that they have a definite stake in the game. These are my judgments as a conservative who is interested in the problem of conservation.

Finally, we have the supremely horrible problem of war, the effects of which, so far as natural resources are concerned, I need not elaborate. Nor shall I discuss the current, deadly international poke game, except to point out that Great Britain has lost her great role as a stabilizer, not because of any political doctrines but because she has become a nation without resources.

Instead I wish merely to point out that, in our own desperate economic plight of the 1930s, we relied heavily on restoring international trade. And aside from scrap iron, oil, and phosphates, we did a tremendous business in selling machine tools to Germany, Russia, and Japan. Machine tools are the real armaments of war. During that period one American company lifted a million dollar deficit, with Japan as its best customer. Another sold to Russia and Germany. I have been told on good authority that these transactions were financed by both British and American capital. You all know what the ensuing war did to our oil, timber, and iron reserves.

I have tried, in the foregoing sketch, to show first that conservation is fundamentally an attitude rather than being simply another technique. Yet that attitude must assuredly employ techniques and direct their use if it is to find effective expression. Of the manifold facets of our present culture, I have selected two for their vital relationship to our problem. One is science, the other private enterprise.

Only by the widest diffusion of scientific understanding can we be brought to realize our obligations as a part of nature. Only when technologists have the kind of scientific training which enables them to grasp the world of nature in its entirety, instead of in closed segments, can we develop the means for making conservation effective.

So far as private enterprise is concerned, its freedom always has been, always must be a relative thing. But its largest measure of continuing freedom can only be won by an honest responsibility for the preservation of our capital structure in terms of natural resources. Not the highest immediate profits but the guarantee of sustained returns must be the goal of our economy.

ORIGINAL PUBLICATION DETAILS

By Paul B. Sears, Head of the Botany Department, Oberlin College, Oberlin, Ohio.

Sears, Paul B. 1950. Conservation—a brief appraisal. *Journal of Soil and Water Conservation* 5(2):55-56,85.

The Country Beautiful

Firman E. Bear

A merica, primeval, undisturbed, in the days of the Indians, was a beautiful place. But much of the virgin forest and soil gave way to the settler with his axe and plow.

Even so, a new beauty has come—in the form of millions of acres of well-managed, productive, colorful domestic crops of fruits, vegetables, cereals and grass. A necessary beauty, to care for the needs of 180 million people.

Along with this though has also come unnecessary ugliness, waste and destruction, in plundered, naked, eroding soil scarred with countless gullies. There is an unfortunate correlation here. Ugliness in natural environment tends to produce ugliness in human behavior and outlook. Slums both in city and country, breed crime, immorality and ignorance.

This goes on, today, unchecked. We are still in a libertine stage of attitude to our natural resources. If a man wants to plow up a hill—that is his license, as far as present laws are concerned, over all but a few isolated situations in this country. If a man wishes to cover the best agricultural land in the nation with pavements, houses, factories, roads or what have you—let his own poor judgment guide him. We're still lacking in legal code and even in politic conscience, as regards an obligation to posterity in this respect.

Interesting experiments have gone on and are still developing to close this abyss. The small watershed approach has been effective in joining local people's interests toward a better view of their conservation obligations to the community. On a larger scale, there has been the Tennessee Valley Authority with its 29 dams for water control and power generation. Similarly Texas has the Brazos River Authority with a program of dams and a chain of lakes 250 miles long. Ohio has its unique Muskingum Conservancy District, a far-flung land and water improvement program stimulated by floods in recent times; plus a similar project in the Miami valley, again spurred by devastating floods that occurred just before the first world war. New Jersey has its Stony Brook project, near Princeton, an excellent example of what a local community can do for itself.

Conservation Starts with Man

A shining goal for all of these programs, aside from physical achievement, is to develop in young and old a high sense of responsibility toward the soil and other renewable resources.

For conservation is a way of life. It starts with man himself. It begins anew each day as he takes his morning bath, shaves, and puts on a clean shirt. It induces him to keep his trousers pressed and his shoes shined. It requires that he make himself presentable, first to himself, then to his family and finally to his fellowmen. He aims to look and act his best. *This is conservation of self respect.*

If he owns a home, the conservation-minded man keeps it painted and in good repair. He strives for an attractive lawn that is free of weeds. He helps his wife in landscaping the premises and he encourages her with her flowers. He tries to put the idea of the orderliness and neatness of Nature across to his children. No trash is permitted to accumulate about his property. Leaves and other plant refuse are made into compost or carted away. Paper and similar wastes are picked up regularly and disposed of. Everything is kept neat and clean about the premises. *This is conservation of home and family.*

If he is a farmer, he sees that something is made to grow wherever a bare spot appears. He will not permit the development of gullies that grow ever larger on his farm. He endeavors to keep the water that flows through his land crystal clear as much of the time as possible. He grows his cultivated crops on the contour, carries off the surplus water in sodded runways, and sows cover crops to protect his land during the winter months. Rough land is seeded down to grass or planted to trees. *This is conservation of soil.*

Such a man takes the necessary measures to insure that as much as possible of the water that falls as rain soaks into the soil where it fell. To this end, he leaves the crop refuse on top of the land wherever feasible instead of plowing it under. He constructs a pond to store water for emergencies to provide a recreation center for his family and friends. He participates in the activities of his soil conservation district in the building of up-stream dams for flood prevention and as a means of lessening the need for down-stream flood control. *This is conservation of water.*

He enjoys the wooded areas, the rugged terrain on which they exist, the tall trees that tower above the surrounding cropland, and the clear streams that flow through them. He works toward their preservation and improvement. He plants more and better trees, and he protects them against damage by animals and fire. *This is conservation of forest lands.*

He is interested in wild animals, birds and fish, and he takes an active part in the development of feeding grounds and sanctuaries to insure their presence in as great abundance and variety as local conditions permit. He encourages his children and those of his neighbors to study the habits of these creatures. And he does not abuse his hunting and fishing privileges. *This is conservation of wildlife.*

If he is an industrialist, he insists that the property under his management be made to look as neat and attractive as the nature of his enterprise permits. If the property is situated within the city, he cleans up the grounds and paints the buildings. If it is located out in the open country, he beautifies the surroundings and develops an attractive recreation center for his employees and their families and builds good will among them and their community. *This is conservation of private enterprise.*

If the processes in his factory are such as evolve dust and smoke, he takes measures to insure that these are reduced to a minimum and prevents the release of toxic fumes. He sees to it that all effluent from his factory is purified before the water is released into a stream. He plans the surroundings to insure sunshine and clean air for his employees. *This is conservation of purity of the atmosphere.*

If he is an architect he seeks to do away with repulsive architecture, dilapidated buildings, automobile cemeteries, weeded areas, and slums, whether in city or country. He even cherishes the hope that the monstrous marble monuments in cemeteries will ultimately be replaced by small flat markers level with the surface of the soil. He wants these sacred grounds to be beautiful places where children can play and old people will not be depressed. *This is conservation of the human spirit.*

If he is a highway engineer, he sees to it that all erodible cuts are protected with vines or shrubs or seeded down to grass. He thinks beyond the two winding strips of concrete that stretch from one great city to the next. He seeks for landscaped space, beautification of the right-of-way, and control of water runoff against possible damage to surrounding lands in times of heavy rains. *This is conservation of public property.*

If he is a teacher, whether in the grades, the high school, the college, the university, the pulpit, or the theater, he bends his energies toward aiding child and man in their quest for knowledge and understanding. He endeavors first to know himself to the end that he can serve better in the uplift of those whom he teaches. *This is conservation of the mind.*

If he is engaged in research, he seeks to discover more about the universe, the Earth on which he lives, and all the other entities with which man is directly or indirectly concerned. He seeks to know more about the life that surrounds him, including himself. But he also seeks to know about the physical and spiritual needs of man. He might even consider research in terms of suitable means by which population growth could be kept under better control. *This is conservation of man himself.*

Conservation Is a Way of Thinking

If soil conservation is this man's profession, his objective is "the promotion and advancement of the science and art of good land use and management to the end that conservation of soil and water and other related renewable natural resources— cultivated crops, livestock, grass, trees, fish, and wildlife—may be useful and enjoyed by mankind forever." Such conservation is fundamental to the continued welfare of the Nation.

Whatever the business or profession of any man may be, there is always need to apply the principles of conservation to his field of endeavor. And there is abundant opportunity and great obligation for him to do something constructive.

Conservation is not only a way of life, it is a way of thinking. And if enough people can be brought to a conservation way of thinking, this can be made and kept the most attractive country on the face of the Earth.

Some of the natural attractiveness of the landscape will have to be sacrificed with further increase in population. But in its place usefulness with beauty can be added in new forms. The great expanses of corn, wheat and cotton can be grown to ever greater perfection. The greatly improved grasses and clovers can come to be an ever greater joy to the eye and mind. The apple and peach orchards to the North and the orange and pecan groves to the South can be made more attractive and fruitful.

There can be many more beautiful public buildings, with spacious grounds. Parks can be enlarged and beautified. Private residences, and the yards and gardens surrounding them can be made more attractive. Finally, there are great possibilities for improvement in the people themselves arising out of a greater appreciation of our renewable natural resources including man himself.

ORIGINAL PUBLICATION DETAILS

Firman E. Bear, Department of Soil Science, Rutgers University, New Brunswick, New Jersey.
Bear, Firman E. 1960. The country beautiful. *Journal of Soil and Water Conservation* 15(4):167-168.

Conservation and the Inexorable Trend of Our Civilization

Paul A. Herbert

S ome sixty years ago a prominent conservationist and chief of a federal bureau pointed out sadly that emotion rather than reason was the basis of many public policies and practices. Probably the more we rely upon argument founded on reason, rather than on emotion, the stronger and more valid will be our position.

In contrast, we often weaken our position when we question the motives of other citizens who disagree with us, and we dissipate our efforts when we set up straw men to tear down or whipping boys to battle.

Then, too, we needlessly antagonize when we clothe ourselves in a sanctimonious façade—that it is selfish to enjoy the tangible or material returns from the natural resources and that it is not selfish to enjoy the intangible and recreational uses of these resources.

In this respect, it is important to remember that the right and the opportunity to earn a living must have priority over the right and the opportunity to enjoy a living. Obviously without the former, we cannot have the latter. I would quickly add, before your hackles rise too high, that this does not mean that one person's opportunity to earn a living should have priority over the opportunity of ten thousand to enjoy a living.

Conservationists, as well as others, are somewhat prone to take the attitude that because they know their position is right, or believe it right, they are justified in using every means to achieve their objective on the theory that the end justifies the means. This is a dangerous and undemocratic philosophy. We can never justify all the means to attain a goal.

We are also inclined to insist that decisions must all be made today—that tomorrow is too late. Then, having established a policy or a program, it cannot and should not be changed for an historic precedent has been established. Obviously conditions are never static, and future generations are more competent than we are to make decisions in their time affecting the conservation of the natural resources.

We generally believe that outdoor recreation, such as hunting and fishing, is necessary to our way of life. Incidentally, there is no scientific proof that the wellbeing of the nation is dependent on any or on all types of outdoor recreation. However, if we assume this is true, then it most certainly is not right to deny the teeming millions in New York City, for example, the equal right to hunt big game in the West, just because the Founding Fathers saw fit to establish certain arbitrary units of government called states. All our people should have a degree of equality to enjoy the recreational product of the nation's resources, regardless of where they live or where the resources are located.

Lastly, in planning long-range conservation programs, it is necessary that we recognize and accept, whether we like it or not, an inexorable trend of our civilization. This nation, as well as others, is rapidly becoming an authoritarian society. This means that the individual liberty of choice is being restricted more and more. This also means that more and more decisions will be made on the federal executive level rather than by the states or in the Congress. We are moving away rapidly from a pure democratic society.

ORIGINAL PUBLICATION DETAILS

Paul A. Herbert, President, National Wildlife Federation.

Herbert, Paul A. 1962. Some thoughts on conservation. *Journal of Soil and Water Conservation* 17(1):27-28.

Public and Private Responsibilities and the Future of Conservation Programs

William Voigt Jr.

T his paper should contain many definitions and boundary markers. When the term "conservation" is unqualified, we take in much resource territory—land, water, vegetation, and the whole animal kingdom. Public responsibilities that touch upon these resources in one way or another could involve every agency in every subdivision of government. Private responsibilities could include every human being capable of the process of reasoning. Between the public and private sectors, there is an overlapping area that is neither one nor the other, but embraces some of both.

Proliferating Responsibilities

Public responsibilities have a tendency to become synonymous with authority and to spread and escalate and proliferate. This tendency has gone so far that there is little likelihood it will be slowed down, certainly not reversed.

To make my point, it is necessary to spotlight what has occurred in one public agency—the Soil Conservation Service. I emphasize at the outset that my reference to that agency does not decrease by one iota my tremendous admiration for it and for the quality and dedication of the personnel employed in it. The Soil Conservation Service compares quite favorably with other resource-oriented federal agencies.

Look back through the years: It doesn't hurt to do this now and then to see where one has been—as long as one does not suffer the delusion that he can go back to what was left behind. The name is the same; the agency is still called the Soil Conservation Service, but nearly all else is different. In the 1930s and 1940s, nearly everyone in the Soil Conservation Service was hard at work forming soil conservation districts, sometimes against stout opposition, enlisting landowners therein as cooperators, and developing basic plans that the Service hoped would later be carried out by those signatories who were conscientious and financially able.

Since then, the responsibilities given to the Service have grown and broadened tremendously. None of these responsibilities was particularly forced upon the agency. The people needed to staff the main office and the field offices of the Service in one of our larger states today is about equal to the number it took to staff the entire Service in earlier times.

District work is still important and properly so. However, in all but the smallest states it is now the responsibility of only one of several assistant administrators. A good many of the assistant administrators are performing duties that were not known when the original Soil Conservation Service charter was written in Kansas dust on congressional desks 30 years ago.

The districts, of course, have become soil and water conservation districts. In addition to strictly district affairs, the Service today is deeply involved in flood control, water pollution control, public and industrial water supply, research, watershed development, economic development, recreation development, fish and wildlife habitat enhancement, forest land management, and still other activities that affect our basic resources of land and water. Another effort is being made this year to increase the permitted size of impoundments built by the Service. This is escalation; the size of the impoundments may one day reach hydropower-producing proportions.

The same burgeoning, the same reaching out—and being pushed out—into fields far from those initially envisioned for them, may be detected in the spans of activities performed by other federal public agencies—one among several in this category being the Corps of Engineers, United States Army. Perhaps, if the Service ever goes into the hydropower business, the Corps will start organizing its own soil and water conservation districts.

The word "integration" in its business sense means an industrial complex that includes, under one management or parent concern, nearly all things needed to produce finished products from raw materials. United States Steel exemplifies that kind of integration in heavy industry. It has its own iron mines, its coal mines for fuel, its limestone quarries and sources of other materials used in processing, its railroads, its fleets of vessels to haul these materials to its steel making mills, its fabricating plants to convert crude or semi-finished steel into products to be offered for sale or erected into structures by its own subsidiaries, and its sales force, warehouses, and other outlets, as well as miscellaneous additional departments.

I use the word in its industrial sense when I say that our resource-oriented public agencies, especially but not exclusively at the federal level, tend toward integration. They seem to welcome all the resource responsibilities it is possible to bring into their organizational orbits. One day this could become total integration, perhaps accomplished in part by assimilation. Then we would have only one massive federal Natural Resources Department, with divisions as large as presently existing departments. Or, there could be two or more blue-sky resource departments with identical congressional charters, competing on the one hand for appropriations and on the other for land and water areas on which to spend the money.

There is a degree of exaggeration in all this. But enough truth can be found in it to make us pause and ask whether we have not given, and are not now giving or showing ourselves prone to give, more than should be turned over to our growing federal establishment. Am I the only one who has a sort of queasy, uneasy feeling about this subject?

Bigness itself, and a degree of integration (as defined) need not and should not be considered wholly bad. My concern goes further.

Standards on the Skids?

We have been trending toward what we are today ever since the founding of the Republic, so perhaps the present situation could be called a form of "manifest destiny." Our ostensible philosophy as a nation has been the doing of the possible. The individual handles problems until they become too much for him, then goes to successively larger subdivisions of government for relief or assistance. Today, we turn from self-help to public aid faster and with less conscientious objection than formerly. There was a big break in the spirit of self-help in the years of the Great Depression. World War II did not check or change the tide. Then came a double decade of steady escalation-integration-proliferation in the resources field. The curve depicting decreasing reliance on self-help rose at a fairly uniform rate; there were not many notable peaks and valleys. But now another huge point of departure, or break in the rhythm, is at hand.

In the after-the-war years, conscientious administrators and lawmakers—and just plain people, too—tried to maintain fairly high standards of excellence for conservation programs and projects. A conscious effort was made to be fairly sure the taxpayer would get something substantial out of his investment.

In order that all of us may one day thrive at a high level of affluence, we now find a vast array of grant, loan, and development programs in operation or in the making. A sizable number of such programs bear the proud name of conservation though resource manipulation might be a more apt designation of what is taking place or is contemplated. This has developed to the point that it may not be brought under control in our lifetimes; the situation may be beyond redemption, even if anyone is in the mood to return to earlier attitudes and actions.

The Appalachia program has much to commend it, and it is close to heresy to question any part of it or anything else intended to abolish poverty. Yet, in our eagerness to help the unfortunate and to raise the economic level of a vast region to that of other parts of the nation, we have started something we may not be able to finish. We have opened the way to a lowering of conservation program standards that is practically beyond comprehension.

A group promoting a Public Law 566 project recently was searching through the grant and loan programs to see how much of the cost would be paid by the generosity of the taxpayer so that as small a part of the outlay as possible would come out of local or private pockets. This seems to be standard procedure. An assistant administrator was heard to say that more and larger grants would be available if the small watershed were located in Appalachia.

In Appalachia, the Corps of Engineers now can give to secondary or indirect benefits status equal to direct benefits in arriving at benefit-cost ratios of water development projects. This means that in Appalachia, tax dollars are used to build reservoirs and other water developments that would have had trouble getting past the first or second cut in comprehensive river basin planning as we have known it.

We have already seen the result of such lowering of standards for Appalachia. Inevitably, it could not be confined to that region. Has it not, in some respects, already burst through those geographic boundaries? Will such lowering of standards not expand until it encompasses the entire nation? Will it not embrace all the natural resources that we in this Society usually think of when we speak of conservation programs?

To sum up, what I have tried to say is this: (a) Instead of examining public and private responsibilities, we should approach from the standpoint of responsibility: there is a difference in the meanings of the two words. (b) We should look critically at the quickened pace of our travel to wherever we may be going. Then, perhaps, we can sit down together and figure out the future of conservation programs.

ORIGINAL PUBLICATION DETAILS

William Voigt Jr., Executive Director, Interstate Advisory Committee on the Susquehanna River Basin, Harrisburg, Pennsylvania.

Voigt, William Jr. 1965. Public and private responsibilities and the future of conservation programs. *Journal of Soil and Water Conservation* 20(4):139-140.

On Stewardship Ethics among Land Managers

James W. Giltmier

Editor's note. This chapter identifies some points of philosophical dissonance within the conservation movement; further discussion can be found in the chapter "Science and Stewardship in a Nonmonolithic Conservation Movement" in part 4.

Many land managers tend to think of themselves as stewards of the land, within the context of shepherding or manipulating the future of the land, for example, or as having dominion over the land. For them, the idea of stewardship is an ethical principle, in these regards:

- For the community (in this case, all of society), scientific stewardship of the land leads to a sustainable future, assuring upcoming generations of guaranteed products and benefits from the land.
- It is also ethical because in caring for the land, the steward provides for the community—more than for himself. This is his or her way to lead what the Greek philosophers called a decent life.
- In terms of economics, sustainable stewardship is good because it ignores short-term profits for long-term, continuing benefits.
- Politically, the land stewards have been able to hold a moral and intellectual dominance in the past because of their personal, physical commitments to the demands of the land and their reliance on scientific principles that were a cut or two above what most citizens understood.

There are, of course, some chinks in the armor of stewardship that have developed of late:

- Because of a large number of ecological and public relations mistakes, public confidence in the ability of land stewards and the infallibility of science has waned considerably.
- New bodies of "experts" have been formed who have redefined the land stewards as the enemy, as people who no longer care about the long-term sustainability of natural resources, and whose only concerns are commodity outputs.
- Some so-called land stewards seem to have gone out of their way to prove that the above allegations are correct.

- Still another batch of persons, who range from those who say mankind is the ultimate destroyer of nature to those who would commit mayhem and murder to preserve resources to their own standards, are successfully instilling in the public a sense of fear. Their basic messages are that governments and their professional land managers are ineffectual and against nature. Science, they say, no longer knows the answers; the only thing for mankind to do is wait for the universe to either implode or explode, while all the right-thinking people live their lives like hermits.

Surely, the people who are suspicious of government land managers have had some legitimate reasons for those suspicions. Without question, science has portrayed itself at times as allknowing, when, in fact, it didn't really know. Often, land stewards have been smug and puffed up about who they are and what they do. Instead of sharing information with an eager public, they have dumped it on the table, then walked away.

Is it unreasonable for the citizens who own the millions of acres of public land to insist that they have a real, not phoney, say about what happens to that land?

At a time of potentially irreparable environmental stress on the globe—a time when we must work together as a community to sustain our human existence here—there is either vast polarization between the major actors on stage, or the rhetorical conflict has made large segments of the public indifferent. They are too busy with day-to-day survival to bother themselves with thoughts about when the apocalypse will come. What in the world can they do about it anyway?

But stewardship aside—as ethical partners with the land, the water, the flora and the fauna—what are the principles that must drive us as we face a series of possible scenarios on the future of our universe? We know that as the population increases, land management must be more and not less intensive. We know that species of plants and animals are being exterminated at terrible rates. We know that there are limits to the use of chemicals to expand crop production for the world. We know that demands for terrestrial commodities will increase as world population increases. If we are not careful, we can reasonably expect that, as the world grows poorer because of the nonsustainable extraction of our resources, it will become more difficult to protect the resources that remain. People who are hungry have little time or concern for environmental protection.

And who will lead in the face of these pressures? Will it be the land managers, the stewards, the seekers after resource sustainability? Will it be those who spike trees? Will it be those who live only for today and the quarterly bottom line? Where are we headed? If things are going badly, what must be done to put them back on track? In a democratic society it may be quite all right from social, scientific, economic, and political standpoints for us to have so many wildly different ideas before us to consider. Then again, if it is true that we have, through our abuse and overuse, brought about the beginning of the end of nature on this little planet, perhaps it is time for us to circle our intellectual wagons to seek some direction and find some leadership.

Those people who have chosen to live their own version of a decent life by dedicating themselves to the stewardship of the land can do one of a couple things. They can sit in a dark corner, unappreciated and defensive, wondering why their attempts to accomplish good were only met with public abuse and indifference; they can hang around waiting for the public to declare them irrelevant to the problems at hand because of their past failures.

Or they can get off their butts and share what they know with the public. They can share the ideas that what is wrong with our globe is not just a scientific or an ethical problem, it is a problem of the community. A lot of decent people have no leadership and don't know what to do. Our land stewards must involve the public in what they need to know to help the professionals save our biological hash. And they must reexamine their communications techniques. Newspapers, for instance, seem to be dying as principal communications media, and the television news networks are not taking up the slack. Their audiences are shrinking as well. Entirely new marketing strategies must be devised.

Heaven knows it will be a lot easier and less contentious to let the "know-nothings" keep the center of the stage than for the ethical conservationists to stand in the pulpit once again as evangelists to sell the conservation movement of Gifford Pinchot and Hugh Hammond Bennett.

But now the stakes may be a little higher. Maybe instead of saving a few inches of Iowa topsoil, our job is going to be to save the world. That's something special to think about. This may be our last and our best ethical chance to lead the decent life that each of us intended for ourselves. Because to paraphrase Yogi Berra, *this time when it's over, it's really over.*

ORIGINAL PUBLICATION DETAILS

James W. Giltmier, Executive Vice President, Pinchot Institute for Conservation, Springfield, Virginia.
Giltmier, James W. 1990. On stewardship ethics among land managers. *Journal of Soil and Water Conservation* 45(6):625-626.

Achieving Public Values on Private Land

R. Neil Sampson

Webster defines value as "the quality or fact of being excellent, useful, or desirable." Few things are more personal or dear to each of us than those things we hold as "basic values." Those values we share with each other within this society, as soil conservation professionals, and in our national cultures as citizens, are counted among our most prized possessions. They are critical, as well, because they form the basis upon which we continually shape and reshape the public policies that form the "rules of the game" under which we all live our lives.

Those values constantly change, for each of us as individuals and for society as a whole. There are individuals reading this who, at one point in their youth, thought that a souped-up Ford was the single most important thing in their life. But we all move on in our lives and adopt different values as we go.

I think it is fair to say that my own personal values, as they relate to land use and resource management, have changed dramatically during my professional career. I suspect that most of you feel the same. For one thing, the information base that each of us has available has expanded so tremendously that as new facts have changed our understanding of values we can't help but have changed.

I was raised on a farm where it had been common for many years to burn wheat stubble after harvest. Burning the stubble, it was said, would kill weed seeds and insect pests, reduce cultivation problems, and help the yields of the following crop. It was a "good" thing to do, my grandfather believed. He and his generation bragged at the local store about the "clean burns" they had achieved.

When I went away to school, agronomy training told me something far different. It told me that burning the crop residue robbed the soil of essential organic matter, released nitrogen because of major changes in the carbon:nitrogen ratio in the soil, and was a prime factor in the depletion of soil organic matter and the excessive winter soil erosion that was occurring. I quickly came to believe that stubble burning was a "bad" thing to do. As you might imagine, that topic was a hot one for a few years around our dinner table and many neighboring farms where Soil Conservation Service (SCS) conservationists were trying to stop this common practice.

When I started working with SCS three decades ago, it was common to assist landowners in "improving" the small streams and creeks that ran through their lands. Streamside trees and brush were taken out, the channel was straightened and sloped neatly, and grassed. We prided ourselves on having helped the farmer develop additional economic land resources, reducing floods, and stopping stream-bank erosion. We were applying the then-current scientific thinking to the challenge of resource management.

Today, landowners and SCS technicians are busy "re-naturalizing" those streams that earlier generations worked so hard to straighten. Log structures, new curves, and riparian tree plantings are all part of the effort to restore more natural watershed regimes. We learned lessons, often at great expense, and a new scientific understanding has emerged as a result.

It is my contention that many of our values have changed, not because we have some elevated sense of morality, but because we know better. We have better information. We understand that we must maintain a better carbon cycle and protect topsoils from erosion if we hope to develop a more sustainable agriculture. So, we work to retain crop residues through new methods of conservation tillage. We know that streams have to have curves, and pools, and riffles, and trees, and sometimes flood out onto the floodplain to function properly.

Private Property Debates

So, as we learn, our values change. But some things, although they change, change very slowly. One example that is critical is the matter of private land ownership rights. Obviously, as we seek a way to gain the recognition of public values on private property, we must deal with this basic issue in a way that works.

The United States has had a continuous, significant debate about the rights of private property since the first Pilgrims hit the shore. That debate is, if anything, more intense today than at any time in recent memory. There are some who say we need to aim more toward the European model, with tighter government control over land use. Others argue fiercely in the opposite direction, saying that increasing government regulation constitutes an illegal "taking" of value without the just compensation called for in the Fifth Amendment.

Behind the rhetoric, however, what is clear is that there are both rights and responsibilities inherent in the ownership of property. Each of us has the right to own and enjoy property, whether that property is a dollar bill, a book, or an acre of farmland. That is an essential right without which a society based upon individual freedoms cannot exist.

But we're not free to use that property in ways that harm our neighbors or that harm society as a whole. We cannot use the dollar to pay for commission of a crime, or the book to hit somebody over the head, or that farmland in a way that diminishes the private or public values of our neighbors.

Even the recent Lucas vs. South Carolina decision of the US Supreme Court, which was looked upon as a chance for a conservative court to force government to stop enacting regulations that might diminish the value of a person's land, was clear in establishing the fact that government still retains significant abilities to regulate land use. What it said, as I read the results, is that if government regulation removes all economically valuable use from the owner, the government should pay compensation. While that sets a new standard for governments to consider, it certainly does not shift the previous balance too far in any one direction.

Keep in mind, however, the Lucas case, the example used by Justice Antonio Scalia to illustrate a time when the person would have no right to compensation was if a lakebed owner were denied a permit to create a landfill if the project might flood nearby land. In other words, if this legal decision is taken in its most narrow context, people cannot do things with their land that might hurt other people around them.

Future Generations and Nonhumans

A good question to ask here, however, is "Which other people?" Is it just our living neighbors that we must avoid damaging? What about the unborn? What rights have they? The entire soil conservation movement, from Hugh Bennett's day until now, has rested heavily on the obligation each of us owes to future generations. Realizing that careless management today can destroy soil productivity for geologic eons, we feel that no person has the right to rob future generations of their essential right to enjoy a productive earth.

The United States has established a national policy that says we should prevent soil abuse that robs our future generations, but then has been willing only to implement that policy up to a point. We're at the stage of giving the idea fairly good lip service, and even legal standing, just like we did 100 years ago with civil rights. But there's a long way to go, just like there was in 1960 with civil rights, before we really address the attitudes and values that people hold.

But while we are still working to gain full societal support for programs that protect the land in the interests of current and future generations of humans, we are faced with the new idea that we should extend our concept of values to the nonhuman portions of the environment. This raises questions that are far more challenging, and difficult.

Some people now propose that land users should save wetlands because they provide a critical link in hydrologic and biological cycles. Those functions have human as well as environmental value, to be sure, but those human values are often indirect and hard to quantify. What we are really saying is something like, "Those swamps aren't worthless just because you can't plow and plant them. They are working very hard in their role as an essential link in the natural environment, and you should protect them because all of society benefits from those environmental contributions and functions."

Whether it is a question of leaving base flow in Western rivers so that aquatic populations can have a chance to survive, or managing a piece of range differently because it harbors an endangered plant or animal, or protecting old growth forest ecosystems to preserve the biological diversity that they alone harbor, or dramatically altering farming practices so that we don't send loads of destructive nutrients into the Chesapeake Bay, today's landowners and resource users are being asked to consider the values of the nonhuman environment and protect those values as part of their land management decisions.

In all of those instances, what we are essentially asking is that values shared by all the public be protected, often at a cost to the individual private landowner or user.

Addressing the Problems

The question then arises: If the public imposes those costs on the private landowner, have they diminished his/her property value to the point where compensation is a fair question? That brings us back to Mr. Lucas, and his appeal for compensation on the basis that the State of South Carolina took away his right to build houses on two lots he bought near the ocean.

On my reading of the Lucas case, the answer is still, "No, except in those rare cases where protecting the public value has caused the loss of all economic use." But Lucas is not the last word. There will be other controversies and other court cases. If nothing else, the so-called "wise use" movement will see to it. And 12 years of conservative national governments have left the United States with a conservative judicial system that will last for decades.

In my judgment, this means only one thing: If today's public wishes to expand the scope of the values it holds important and wishes to ask landowners to limit economic use of property to achieve those values, we need to look at tools other than straight regulation as the way to accomplish the task.

I am convinced that this can be done if we address these new problems in much the same way that we have historically addressed other important resource questions, including soil conservation. Not that we have accomplished everything we need in terms of soil conservation. But we have made enormous progress, and that progress demonstrates, I think, ways to think about these new issues.

Ecosystem Management and Values

I've spent a lot of time recently thinking about "ecosystem management," a term that has been adopted by the Forest Service to describe a land management style based on trying to understand and protect the processes that make an ecosystem function. Other agencies have similar terms—SCS is using "total resource management," for example—to explain a concept that is about the same.

Concepts such as ecosystem management are essential to our expanded view of values because they help identify practical methods to manage natural resources in a way that extends our responsibility beyond today's neighbors, into tomorrow's humans and into protecting the integrity of the environment for its own sake. If your land management system helps keep ecosystems healthy and functioning normally, your chances of achieving protection for all values—present and future, human and nonhuman—are as good as today's science knows how to achieve.

But it is one thing to have the Forest Service trying to figure out how to do this on a national forest and quite another for soil conservationists to try to figure out how they are going to get thousands of private landowners to do it on their properties.

Don't get me wrong. I'm not sure that having every landowner do the same thing is a good idea, even if it were possible. One of my Southern friends recently remarked in a conversation on land management aimed at achieving biodiversity that having "several thousand landowners all doing their own thing on their own land was a form of diversity that we should not discount." I agree. There is great value in having large numbers of people acting independently. The result is a "whole" that is both diverse and strong.

But what do we do when the prevailing value system among those large numbers is ignoring an important value in the eyes of the total public? That is where we seem to be heading today. The answer is that we need to find a way to change the prevailing value system without losing the strength and diversity that comes with a system based on private property rights and individual freedoms.

Historical Perspective

We have a lot of history to tell us how to proceed. In the late 1920s, when it was a common practice for people to set the woods on fire in the South, my organization, then called the American Forestry Association, became concerned. Its foresters were convinced that good forest management in the region would never be possible until fire damage was reduced. So AFA raised the money, bought some trucks, mounted portable generators and movie projectors on them, and sent them into the South to begin teaching the damage in wildfire. Called the "Dixie Crusaders," these roving foresters showed thousands of rural southerners their first moving pictures, projected on the wall of the general store, preaching the gospel of fire control.

In one Idaho soil conservation district where wheat stubble burning was rampant and conservationists were doing everything to stop it, a series of roadside jingles, in the old "Burma Shave" mode, did more to end burning than any other effort. When a local fire made its perpetrator the butt of jokes at the coffee shop instead of a hero, burning came to a quick end.

Lessons to be Learned

The first lesson, then, is we must educate ourselves, perhaps first and most of all. How many resource professionals today feel fully competent in discussing such concepts as ecosystem structure, composition, and function? How many truly feel that, if they could make the land management decisions themselves, they would know exactly what to do, and when? When they tell a landowner, "Do this, and that, and everything will be all right," how confident are they?

For most of us, learning only makes us appreciate how much we don't know. Our confidence is tempered by the certain knowledge that new information, feedback, and insight is almost certain to alter our most closely-held views. We must educate landowners, but not with simplistic notions. Ecosystem management—or total resource management—is not easy, and it is certainly not some "return to the old ways." It takes all the science and skill we possess, and all that we can learn, as our experience and experiments proceed.

The second lesson is that people must cooperate. It is the rare land manager who can control a complete ecosystem. Even then, there are constant impacts from surrounding ecosystems. Every manager, no matter what the size of the land unit, has neighbors. And what one does affects the other. No person—and no land—is an island, in the sense that it functions alone.

Cooperation and Land Values

Cooperation isn't new. Western landowners long ago set up water districts because it was the only civil way they could share a limited resource. Soil conservation districts, and the entire soil conservation movement, are based on the idea of cooperation. But we may have to take this kind of cooperation to new levels. Let me give you an example from forestry.

In many public forests there are land ownership patterns best described as "checkerboard." They date back to the days when railroads were granted alternate sections of public land as an inducement to build the railroads linking East and West. Today, that results in a map where, if the public land is green and the private land is white, large areas look exactly like a checkerboard.

Enter the public land management agency, let's say the Forest Service. It is studying a watershed and trying to establish an acceptable amount of disturbance in that watershed—one that will not upset the functioning of the system. The best scientific estimate is that, in that watershed, not over five percent of the land should be harvested for timber in any one year. So, it sets up a harvest plan that is consistent with the sustainable yield of the forest and that keeps the harvest impact on the watershed within limits.

But along come the private landowners, who decide that prices are right to sell off the timber value they have been holding for many years. They cut a bunch of their land, and suddenly instead of a five percent impact on the watershed, they have created a 20 percent impact.

What is the Forest Service to do? Must they sit back and void all their plans and wait until the private neighbors finish up and the watershed is back into balance before they go ahead with their own management plans? What other choice do they have?

Cooperative Management

Clearly, we need a new form of cooperation in such cases. Maybe we could devise some sort of "management cooperative," in which each landowner agrees to follow a set of management guidelines that have been jointly developed and agreed upon.

It is possible that we could couple this "management cooperative" with the concept of marketable impact permits. Under the new Clean Air Act, industry is testing the idea of emissions trading. Maybe a watershed full of livestock operations somewhere in the Chesapeake Bay region can establish an acceptable level of nutrient loss that is shared by all owners. An owner who for some reason could not meet his or her allotted release level without unusual cost could buy unused permits from a neighbor who, for reasons associated with his or her operation, found that he or she had more permits than needed. The beauty of this idea is that it leaves great latitude for individual choice, lets all parties seek the most economic solution in a free-market trading climate, and still achieves the goal of limiting pollution to an allowable limit.

This same kind of idea might work for forest landowners in a watershed, where "timber harvest permits" could be given out in a fair proportion, then traded back and forth in a free market to help meet each owner's individual situation and needs. That might get at the "checkerboard" problem.

One major problem with this idea is the inability of public land managers to make management decisions as freely as their private neighbors. Too often, they are restricted by legal regulations, budget limits, or management targets set at higher levels. If, for example, all the parties agree that the public agency should make certain improvements on their land to meet the joint need, that does nothing to change the national or regional budget, and that money may simply not be available. Ranchers in the West who have tried to get public neighbors to control weeds, stop excessive runoff from highlands onto downstream private land, or fulfill their part of joint agreements on fencing and livestock water development know this problem all too well.

Public and Private Cooperation

This means that ecosystem management in mixed public-private ownership may demand some changes in the way public land agencies operate, as well as requiring private landowners to join and participate.

The question, then, is, "How do we get this kind of thing started?"

Again, I would argue that we have the models in place. It is just a matter of continuing to develop and refine them. We have models of public-private cooperative planning on rangelands in the West. We have developed estimates of allowable pollution loadings for streams in critical watersheds. We have many examples of private covenants that guide land use and management in the urban and suburban areas of the country. We have conservation easements that allow the public to purchase private rights for fair market value, and emissions trading that allows private interests to seek a market solution to the challenge of reducing pollution.

What we have not done, as yet, is bring these models into widespread usage to help us cooperate with each other on the management challenges we will face when we begin to insist, as a society, that landowners improve their ability to better protect the inherent integrity of the ecosystems they manage.

Final Analysis

Before we can take that step, it seems to me we have a significant way to go in educating ourselves, as resource professionals, on the costs and benefits associated with ecosystem management and on the practical methods of achieving it. We need to have become convinced that this is a highly valuable goal and be able to articulate why it is valuable in ways that the public understands and comes to agree upon.

In the final analysis, our ability to achieve public values on private land will not come about simply because the nation passes laws restricting pollution or encouraging soil erosion control or protecting endangered species. We will only begin to achieve these goals when we have converted them into widely shared social values that people understand and agree upon.

And, in perhaps the most challenging aspect of all, we will only begin to achieve public values on private land when you and I, as resource professionals, agree on the fact that these are values that we hold in high esteem. In addition, then, we must also have done the scientific research and testing so that we can say with confidence that we know how to help private landowners in their search for practical ways to achieve these public values.

That is, it seems to me, the ultimate value of a professional society like the Soil and Water Conservation Society. Let each of us be determined to participate to the fullest, to gain a broader set of visions, a more complete sense of the values we share, and a new enthusiasm for applying those shared values to the daily tasks we face.

ORIGINAL PUBLICATION DETAILS

R. Neil Sampson, Executive Vice President, American Forests, Washington, DC.

Sampson, R. Neil. 1992. Achieving public values on private land. *Journal of Soil and Water Conservation* 47(5):360-364.

Putting "Ecosystem" into Natural Resource Management

Fred B. Samson and Fritz L. Knopf

Many resource managers are excited about the possibility of using an ecosystem approach for real-world applications in resource management (Sparrowe 1994). Ecosystem management is pivotal to the growing national debate over the best use and management of public lands and the relative worth society places on the commodities and social values those lands provide. Similar concerns about the management of private lands have been voiced, particularly over the need to balance economic issues with the national mandate to conserve species thought to be threatened or faced with extinction (Clark 1996). Clearly, the traditional way America's natural resources have been managed does not protect their natural values (Sax 1993) or provide for the sustainable production of goods and services (Barnes 1993).

Leopold described ecosystem management in two distinct stages; first, determine ecological limits inherent to an ecosystem and, second, develop resource use plans within those ecologically based ecosystem constraints (Norton 1991) with the initial involvement of the public, conservation groups, and agencies. However, available data do not allow for the assessment of ecological limits inherent to most if not all ecosystems (Beier and Rasmussen 1994). New models—the Great Plains Partnership, the Joint Venture notion, and the Forest Ecosystem Management and Assessment Team model—to describe and conserve ecosystems are proposed, but no single, widely agreed upon approach to develop ecosystem-level resource-use plans has evolved (Samson and Knopf 1994). This inconsistency only serves to confuse the public, policy makers, and managers, impeding efforts to implement ecosystem management.

This paper describes an approach to ecosystem management that involves four distinct steps: (1) a model of technical principles and concepts in ecosystem management; (2) a method to monitor the precise relation between land management activities that impact an ecosystem and the ecosystem itself; (3) an approach to manage and or restore an ecosystem founded on principles of ecosystem dynamics; and (4) a concept to develop the vision and strategy to manage an ecosystem. We conclude with a focus to put "ecosystem" into natural resource management.

An Ecosystem Model

Holling uncovered four primary phases of an ecosystem cycle: exploitation, conservation, release, and reorganization. Exploitation is the rapid colonization by plants, animals, and microorganisms of recently disturbed areas. Conservation is the accumulation and storage of energy following the establishment phase. In release, the bound accumulation of nutrients and biomass is suddenly released by agents such as insect pests, forest fires, intense and frequent grazing, or a combination thereof. The release of nutrients and biomass is then reorganized to permit renewal of the same or "natural" ecosystem. Reorganization, the fourth phase, is the capture and cycling of nutrients, energy, and water within the area to become available for the next phase of exploitation and to permit ecosystem renewal.

The potential of an ecosystem to produce commodities and satisfy social values depends on the amount of energy captured and on the accumulation and cycling of nutrients over time (National Research Council 1994). The critical consideration in ecosystem management is whether extrinsic disturbance or human management disturbance of the accumulated biomass and nutrients over too large an area results in a permanent loss of ecosystem capacity to produce commodities and satisfy social values. Further, conservation and management of natural communities and ecosystems to produce commodities and social values may be impossible without maintenance of natural processes (Smith et al. 1993). Thus, the protection and health of these natural processes should be the central goal of ecosystem management.

The link between a prevailing ecosystem cycle and an altered ecosystem is defined by four factors: (1) The disruption of an ecosystem is not easily reversed, at least within sensible timeframes and costs. (2) There is an increase in non-indigenous species—those plants, animals, and microbes found beyond their natural geographic range. (3) Non-indigenous species affect ecosystem-level properties, i.e., productivity, nutrient cycling, and hydrology. (4) A more invasive pattern in type and extent of ecosystem disturbance may evolve to the extent that the survival of society as we know it is undermined.

The transition from one ecosystem to another may occur by rare yet natural episodic events, such as drought, shift in rainfall pattern, etc., or the cumulative effects of one or more of the ecological stresses. Human intervention may also cause shifts.

The intentional introduction of species for sport, biological control, pharmaceutical material, or ornamental use has had a profound effect in altering natural ecosystems, in contributing to the listing of 160 species as threatened or endangered under the Endangered Species Act (Bean 1991), and is affecting many national interests including agriculture, industry, and human health (Culotta 1991).

Two essential features of a landscape are the composition of habitats and the spatial arrangements of those habitats. Processes that shape the composition and arrangement of habitats important at one scale are frequently not meaningful or predictive at another scale (Gipson et al. 1993). For example, at the local level, the "intermediate" disturbance model predicts that the most common frequency in disturbance coincides with the life histories of the greatest number of plant species (Clark 1991). On a larger or landscape scale, the disturbance heterogeneity hypothesis predicts the number of plant species as related to the extent and timing of disturbance across the landscape (Collins 1992). Such models should serve as an early warning of a shift to a different plant community as the frequency and extent of natural processes are altered. There is mounting evidence that not only the occurrence of disturbances but also the spatial structure of disturbed patches is important to the distributions and abundances of animal species (Hansen et al. 1993).

Together, it is essential to understand the role ecological processes play in structuring and maintaining ecosystems and in the development of a model for ecosystem management. The basic reasons to emphasize the role of ecological processes are, first, that a given land management activity may cause perturbations to an ecosystem that have unintended, indirect effects on other system elements and second, that the direct and or cumulative effects of such activities may cause the demise of ecosystems on whose services humans depend (Dailey and Ehrlich 1992). The delay between predicting and detecting irreversible and deleterious ecosystem-level changes with certainty often delays the receptivity to acknowledging environmental problems and seeking solutions. This uncertainty is compounded over the period of delay; the longer the period, the larger the gamble (Ludwig 1993).

Ecosystem Monitoring
Periodic monitoring is a fundamental part of ecosystem management. Successful ecosystem monitoring (1) provides an early warning of a change in one ecosystem to another; (2) relates land management activities to the ecosystem itself in both the short and long term, and (3) indicates change in the capacity of the ecosystem to produce commodities and satisfy values.

The ecosystem model suggests that the pathways of an ecosystem cycle are circular and simple. This is not the case. Holling suggests that a small set of ecological processes structure ecosystems into a small number of levels; that, in a similar vein, the ecological process that influence ecosystem structure fall into groupings, i.e., processes operate in the microscale of centimeters and days, the mesoscale of tens of meters to kilometers and years, or the macroscale extreme described in thousands of kilometers and centuries; that one cannot simply aggregate fine-scale knowledge given the non-linear nature of ecosystems but knowledge at all levels is required; and that morphological and behavioral attributes of animals can be used to monitor existing landscape structure and to predict change.

Holling proposed that the discontinuous nature of the ecological processes that structure an ecosystem should, in turn, generate a discontinuous distribution in animals. Thus, animals should exhibit attributes of body size and or behavior scaled to the landscape or level of the ecosystem in which they live. Two alternatives were tested by Holling with the conclusion being that birds and mammals show a number of body-mass clumps entrained by the discontinuous and hierarchical structure of an ecosystem independent of either tropic or taxonomic status.

A primary impact of humans on an ecosystem is to rescale in space and time landscape patterns such as size, shape, composition, and juxtaposition of habitats, communities, and ecosystems (Motzkin et al. 1993), ecological processes such as energy flow (Tilman and Downing 1994), nutrient cycling, hydrology (Kinsolving and Bain 1993) and so on, as well as the introduction of non-indigenous species (Vitousek 1990). Though these effects are not independent, the specific actions of rescaling in land and resource management impact commodity and social values of ecosystems.

The relationships between scale-specific actions and scale-dependent distributions of animals enables a scale-dependent approach to ecosystem monitoring. Ecological processes at each ecosystem level have a characteristic natural frequency and, typically, a corresponding temporal and spatial scale. We chose to select an "indicator" species from each body-mass clump based on home range. The presumption being large home range and body size are thought to increase a species vulnerability to ecosystem change and or extinction. The ability to measure native species' constancy at differing levels within an ecosystem is important; constancy in species presence is viewed as evidence of a "normal" ecosystem; change in species constancy implies a shift in ecosystem function (Angermeir and Karr 1994).

Ecosystem Management

We can greatly increase the odds of making environmentally sound ecosystem management decisions at the ecosystem to site level by following several steps:

- Recognize the ecosystem as the resource (Odum 1992).
- Recognize that all species are not created equal, some are extinction-prone (Sisk et al. 1994), particularly the narrow endemic species that occur in one major or macrohabitat (Mares 1992).
- Emulate and tailor to scale the process-dependent structural and functional attributes of an ecosystem in land management activities (Gipson et al. 1993).
- Emphasize methods to manage ecosystems to answer many different questions: social, economic, and ecological (Costanza et al. 1993).

Samson et al. (1991, 1992) provided the first published reference on the utility of near ecosystem-wide resource-use plans. They wrote about one of the most unique ecosystems found in North America, Southeast Alaska's temperate rain forest. Most (–95%) of the rain forest is part of the USDA Forest Service Tongass National Forest. The approach taken in Southeast Alaska was one of scale with a full understanding of the ecological processes operating within this unique environment. First, the ecological units that in a general way describe ecosystems were defined, specifically the old-growth temperate rain forest plant communities. Second, a province system was established that captures representative samples of plant communities and, therefore, habitats for all species dependent or closely associated with that habitat. Third, it was recognized that speciation since the last glacial period creates a significant management issue in conservation—the extinction-prone narrow endemic.

In Southeast Alaska, the temporal and architectural structure of the temperate rain forest is in general determined by two classes of wind-dependent ecological processes: low intensity but very frequent individual tree windthrow creating small gaps (2 to 4 ha) in the forest canopy, and high intensity but infrequent windthrow creating large (up to 200 ha) openings. The fourth step was to recommend that timber harvest emulate natural patch dynamics while giving consideration to other land-use requirements, including the visual quality of the landscape, protecting against siltation of streams to conserve anadromous fish resources, and the economics and techniques of timber harvest; and that at least one watershed be left intact, since maintenance of an entire watershed would maintain viable populations of most, if not all species in the short term. Over the long term, landscape elements (i.e., the size, shape, and distribution of habitats, in those watersheds entered for commodity production) will revert to a natural condition. Likewise, the watershed initially designated for maintenance of old-growth dependent species will, at some point in the future, be entered for commodity production.

The traditional or "press" approach in forest resource management, where each watershed is entered, differs from the "pulse" approach recommended in Southeast Alaska. Ecosystems and their component landscapes are dynamically changing habitat mosaics. Attempts to perpetuate a landscape or ecosystem at the watershed level present a nearly insurmountable problem. For example, the agents of vegetation change, such as fire and windthrow, cross watershed borders with ease and often with no predictable pattern (Johnson and Larsen 1991) and preclude a watershed-by-watershed approach to long-term, ecosystem wide planning. The "pulse" or ecosystem approach to resource management allows saving an ecosystem as a regional resource rather than treating each management activity as a unique environmental entry, prioritization and harvest of resources in a prudent and timely manner, and protection of social, economic, and ecological values over time.

Vision and Strategy: An Example

Ecosystem management—"people doing something bounded in space" (Gordon 1993)—differs from traditional resource management in three ways. First, the health of both landscape- and ecosystem-level processes that support and sustain species, communities, and ecosystems over time is a priority. Second, success is dependent on the selection of the right ecological factors—ecological processes and extinction prone species and so on—rather than on heavy investments of labor and materials. And third, the connections among social values, economic resources, and ecological factors must evolve with understanding (Samson and Knopf 1996a).

Successful management strategies are based on vision, not planning (Mintzberg 1994). Broad visionary conservation goals, identified in the 29 federal laws that relate to the conservation of biological diversity and therefore to ecosystem management, fail to provide an explicit framework to conserve the variety of life and ecological processes characteristic to a particular ecosystem, and, thus, are of limited value in ecosystem planning and management. A "sense of place" is described by the natural history, other descriptive sciences, and the people who live within and recognize the realities of living-in-place (Flores 1994). Many scientists and conservationists argue that diverse, locally managed ecosystems both protect biological diversity and effectively sustain the use of natural resources (Andrade and Rubio-Torgler 1994; Salasfsky et al. 1993).

The Great Plains of North America has a "sense of place" (Flores 1996)—grasslands, history, ways of life, quality of life, resources, institutions, communications, economies, and governance. Nonetheless, nearly 60 percent of the Great Plains area lost rural, small-town populations in the last decade and that out-migration and economic decline is expected to continue. However, the potential for species extinction on grasslands is of serious concern. Fifty-five species in the United States are threatened or endangered; grassland birds have shown more consistent and steeper geographically widespread declines than any other grouping of North American birds; several species no longer exist; and an exotic, crested wheatgrass is among serious economic and ecological threats to the ecosystem (Samson and Knopf 1996b).

The solution is an ecosystem-wide conservation strategy, the Great Plains Partnership (Clark 1996). Implementing the strategy requires a fundamental shift to the sense of place—people who live and work on the Plains are given the opportunity to define and create their own generationally sustainable future and, at the same time, to preserve and enhance the biological uniqueness of the prairies, wetlands, and waterways (Nelson 1996). The goal is to protect and strengthen both the economic and ecological interests through a cooperative effort involving all levels of government, environmental groups, and farming and ranching associations (Mack 1996). With more than 95 percent of the land in private ownership, the possibility of sharp conflicts over threatened and endangered species protection is high. The resulting economic disruption, resentment towards government, and divisiveness would cause major social damage. Solutions will be developed by local communities, but they can seek assistance in removing institutional barriers, obtaining resources, and developing appropriate science (Knopf and Samson 1996).

Summary

Ecosystem management is not a new concept to many in state and federal resource management, and many able resource managers are concerned with the rate and fervor in which ecosystem management is being thrust upon agencies and the public. The reasons are many, ranging from an arrogant attitude as if some advocates have a new idea (Sparrowe 1994) to the suggestion that large levels of natural variability in ecosystems may mask effects of over-exploitation of natural resources (Ludwig et al. 1993). Many resource efforts stall or fail because of improperly handled controversy stemming from differing perspectives, information gaps, and or misunderstandings.

The immediate tasks before ecosystem resource managers and scientists for putting "ecosystem" into natural resource management are as follows:

To understand more fully the history of ecosystem management—the land ethic, hierarchical framework, and selection of the right ecological factors to manage ecosystems effectively and efficiently (Leopold 1933). The current understanding— a relatively small set of plant, animal, and abiotic processes structure ecosystems across scales in space and time (Holling 1992)—serves to strengthen the tie to history, the need to prevent misuse and consequent loss of existing ecosystem resources, and to guide strong and diverse opinions on ecosystem management toward consensus.

To increase the understanding of the consequences of ecosystem change to the extent that ecosystem monitoring is the norm (National Research Council 1994). Monitoring the consequences of ecosystem change faces two fundamental issues; conveying to the public what monitoring of ecosystem change is worth—fire, insect outbreak, plant disease, water flow, and so on (Dailey and Ehrlich 1992): and second, given that the first signs of environmental stress usually occur at the population level (Odum 1992), to select species which are relatively easy and cost effective to monitor and provide for an early warning to changes in ecosystem processes and composition (Holling 1992).

Given the importance of ecosystem management, surprisingly little space in the literature has been devoted to explaining the real-world side of implementation (Morah 1994). The shift in conservation to broad-scale ecosystem management has given rise to three key ecological issues in implementation: the integration of pattern and process in the landscape management activities, use of species to monitor ecosystem change, and "pulse" landscape management activities. The goal is to establish a broad-based ecosystem management strategy for natural resource conservation, based on ecological areas and priorities rather than unique entries into a natural ecosystem, in a cost-efficient manner.

Understanding a community and its role within the larger regional culture is a basic step toward the integration of ecology, economy, and people (Snyder 1990). The demands of rapidly expanding human population require continued economic development to meet basic needs and will likely erode stocks of already scarce natural resources (Costanza et al. 1993). The challenge is the synthesis of technology and cultural practices and values—a sense of place—as it aims to save both the dynamics and structure of ecosystems and achieve the sustainable and equitable use of natural resources. Leopold anticipated the solution when he wrote the "best" plan is one mutually satisfactory to all factions—ecological, economic, and social.

The mechanics of ecosystem management demand a different paradigm in natural resource conservation. The new paradigm must view public agencies as partners or facilitators (rather than regulators) working to promote and protect the uniqueness found in those varied places where peoples choose to live. Ultimately, regulation must come from local societal pressures. Understanding and opening dialogues in cooperative ventures moving toward a conservation vision are the foundation of effective ecosystem management. Just as natural resources are lost with ecosystem disfunction, so too are the capabilities of these systems to support economies and societies.

References

Andrade, G. L., and H. Rubio-Torgler. 1994. Sustainable use of the tropical rain forest: evidence from the avifauna in a shifting-cultivation habitat mosaic in the Columbian Amazon. Conservation Biology 8:545-554.

Angermeier, P.L., and J.R. Karr. 1994. Biological integrity versus biological diversity as policy directives. BioScience 44:690-697.

Barnes, R.L. 1993. The U.C.C.'s insidious preference for agronomy over ecology in farm leading decisions. University of Colorado Law Review 64:457-512.

Bean, M. J. 1991. The role of the U.S. Department of Interior in non-indigenous species issues. Report for the U.S. Congress Office of Technology Assessment, Washington, DC.

Beier, C., and L. Rasmussen. 1994. Effects of whole ecosystem manipulations on ecosystem internal processes. Trends in Ecology and Evolution 9:218-223.

Clark, J.S. 1991. Disturbance and population structure on the shifting mosaic landscape. Ecology 72:1119-1137.

Clark, J.S. 1996. The Great Plains partnership. *In* Prairie Conservation: Preserving North America's Most Endangered Ecosystem, ed. F.B. Samson and F.L. Knopf. Covelo, CA: Island Press.

Collins, S.L. 1992. Fire frequency and community structure in tallgrass prairie vegetation. Ecology 73:2001-2006.

Costanza, R., L. Wainger, C. Folke, and K. Maler. 1993. Modeling complex ecological economic systems. BioScience 43:545-555.

Culotta, E. 1991. Biological immigrants under fire. Science 254:1444-1449.

Dailey, G.C., and P.R. Ehrlich. 1992. Population, sustainability, and earth's carrying capacity. BioScience 42:761-771.

Flores, D. 1994. Place: an argument for bioregional history. Environmental History Review 18:1-18.

Flores, D. 1996. A long love affair with uncommon country: environmental history and the Great Plains. *In* Prairie Conservation: Preserving North America's Most Endangered Ecosystem, ed. F. B. Samson and F. L. Knopf. Covelo, CA: Island Press.

Gipson, D.J., T.R. Seastedt, and J.M. Briggs. 1993. Management practices in tallgrass prairie: Large- and small-scale experimental effects on species composition. Journal of Applied Ecology 30:247-255.

Gordon, J.C. 1993. Ecosystem management. An idiosyncratic overview. *In* Defining Sustainable Forestry, ed. G. H. Aplet, N. Johnson, J. T. Olsen, and V. A. Sample, 240-246. Covelo, CA: Island Press.

Hansen, A.J., S.L. Garman, B. Marks, and D.L. Urban. 1993. An approach for managing vertebrate diversity across multiple-use landscapes. Ecological Applications 3:481-496.

Hobbs, R.J. 1991. Disturbance as a precursor to weed invasion in native vegetation. Plant Protection Quarterly 6:99-104.

Holling, C.S. 1992. Cross-scale morphology, geometry, and dynamics of ecosystems. Ecological Monographs 62:447-502.

Johnson, E.A., and C.P.S. Larsen. 1991. Climatically induced change in fire frequency in the southern Canadian Rockies. Ecology 72:194-201.

Kinsolving, A.D., and M.B. Bain. 1993. Fish assemblage recovery along a riverine disturbance gradient. Ecological Applications 3:531-544.

Knopf, F.L., and F.B. Samson. 1996. Ecology and Conservation of Great Plains vertebrates. New York: Springer Verlag.

Leopold, A. 1933. Game Management. New York: Charles Schriber.

Ludwig, D., R. Hilborn, and C. Waters. 1993. Uncertainty, resource exploitation, and conservation: Lessons from history. Science 260:17,36.

Mack, G. D. 1996. The Sandhill management plan: A partnership initiative. *In* Prairie Conservation: Preserving North America's Most Endangered Ecosystem, ed. F. B. Samson and F. L. Knopf. Covelo, CA: Island Press.

Mares, M. A. 1992. Neotropical mammals and the myth of Amazonian biodiversity. Science 255:976-980.

Mintzberg, H. 1994. The fall and rise of strategic planning. Harvard Business Review (January-February):107-113.

Morah, V. 1994. A different view of sustainability. Ecological Applications 4:405-407.

Motzkin, G., W.A. Paterson III, and N.E.R. Drake. 1993. Fire history and vegetation dynamics of a Chamaecyparis thyoides wetland on Cape Cod, Massachusetts. Journal of Ecology 81:391-402.

National Research Council. 1994. Rangeland health. New methods to classify, inventory, and monitor rangelands., Washington, DC: National Academy Press.

Nelson, E.B. 1996. Foreword. *In* Prairie Conservation: Preserving North America's Most Endangered Ecosystem, ed. F.B. Samson and F.L. Knopf. Covelo, CA: Island Press.

Norton, B.G. 1991. Context and hierarchy in Aldo Leopold's theory of environmental management. Ecological Economics 2:119-127.

Odum, E.P. 1992. Great ideas in ecology for the 1990s. BioScience 42:542-545.

Salasfsky, N., B.L. Dugelby, and J.W. Terborgh. 1993. Can extractive reserves save the rain forest? An ecological and socioeconomic comparison of nontimber forest product extraction systems in Pettn, Guatemala, and West Kalimantan, Indonesia. Conservation Biology 7:39-52.

Samson, F.B., G.C. Iverson, R.M. Strauss and J.C. Capp. 1991. New perspectives in Alaska Forest management. Transactions of the North American Wildlife and Natural Resources Conference 56:652-661.

Samson, F.B., P. Alabach, J. Christner, T. DeMeo, A. Doyle, J. Martin, J. McKibben, M. Orme, L. Surring, K. Thompson, B.G. Wilson, D. A. Anderson, R.W. Flynn, J.W. Schoen, L.G. Shea, and J.F. Franklin. 1992. Conservation of rain forests in Southeast Alaska: Report of a working group. *In* Wildlife and Habitats in Managed Landscapes, ed. E. Rodiek and E. G. Bolen, 97-113. Covelo, CA: Island Press.

Samson, F.B., and F.L. Knopf. 1994. A framework to conserve biological diversity through sustainable land management. Transactions of the North American Wildlife and Natural Resources Conference 59:367-377.

Samson, F.B., and F.L. Knopf. 1996a. Ecosystem management: Selected readings. New York: Springer-Verlag.

Samson, F.B., and F.L. Knopf. 1996b. Prairie conservation: protecting North America's most endangered ecosystem. Covelo, CA: Island Press.

Sax, J.L. 1993. Nature and habitat conservation and protection in the United States. Ecology Law Quarterly 20:47-56.

Sisk, T.D., A.E. Launer, K.R. Switky, and P.R. Ehrlich. 1994. Identifying extinction threats. BioScience 44:592-604.

Smith, T.B., M.W. Bruford, and R.K. Wayne. 1993. The preservation of process: the missing element of conservation programs. Biodiversity Letters 1:164-167.

Snyder, G. 1990. The Practice of the Wild. San Francisco, CA: North Point Press.

Sparrowe, R.D. 1994. Opening statement. Transactions of the North American Wildlife and Natural Resources Conference 59:1-6.

Tilman, D., and J. A. Downing. 1994. Biodiversity and stability in grasslands. Nature 367:363-365.

US Congress Office of Technology Assessment. 1993. Harmful non-indigenous species in the United States. Washington, DC: Government Printing Office.

Vitousek, P.M. 1990. Biological invasions and ecological processes: toward an integration of population biology and ecosystem studies. Oikos 57:7-13.

ORIGINAL PUBLICATION DETAILS

Fred B. Samson, Terrestrial Ecology and Assessments, USDA Forest Service, Missoula, Montana. Fritz L. Knopf, Vertebrate Ecology, USDI National Biological Survey Midcontinental Research Center, Fort Collins, Colorado.

Samson, Fred B., and Fritz L. Knopf. 1996. Putting "ecosystem" into natural resource management. *Journal of Soil and Water Conservation* 51(4):288-292.

Soil and Water Conservation Is Essential for Ecosystem Management

Tony Prato

Public relations experts tell us that attention-grabbing metaphors go a long way toward advancing public support for and understanding of new concepts and ideas.

Public support for soil conservation in the 1930s was greatly enhanced by the Dust Bowl metaphor. The devastation of the Dust Bowl spawned the Soil Erosion Service and its eventual successor, the Natural Resources Conservation Service. Most conservationists would agree that Aldo Leopold's "land ethic" was a metaphor that had profound and long-lasting impacts on the conservation movement.

When, over 10 years ago, managers of Yellowstone National Park decided to let self-governing natural processes play a greater role in the park, they chose to label their new management philosophy as "natural regulation." In recent times, a relatively new metaphor has been used to usher in a new philosophy of resource conservation and management, called ecosystem management (EM). I would like to describe this metaphor and its implications for soil and water conservation.

EM emphasizes large spatial scales, longer time periods and many variables (Thomas 1997). While EM is applicable to both public (state and federal) and private lands, public land managers and environmental groups are its strongest proponents. Eighteen federal agencies have adopted or are considering adoption of programs based on an ecosystem approach to land and water resource management (Haeuber and Franklin 1996).

At the heart of EM is the philosophy that natural resources should be managed in a way that balances social, economic, and environmental values (MacKenzie 1996). EM envisions a two-way relationship between human beings and nature, a concept that can be traced back to the writings of George P. Marsh (1801–1882), particularly his influential book *Man and Nature*. The two-way relationship recognizes that human activities have ecological impacts, such as nonpoint source pollution and loss of biodiversity; ecosystems provide valuable services, such as air and water purification, mitigation of floods and drought, detoxification and decomposition of wastes, generation and renewal of soil, maintenance of biodiversity and partial stabilization of climate (Daily 1997).

Widespread embracing of EM suggests a shift in the philosophical basis of resource management from food, fiber, and forage production to protection and restoration of ecosystems (National Research Council 1992; Williams et al. 1997). Diaz and Bell

(1997) point out that "on federal lands (in the United States) the concept of resource management (in the sense of managing the production of individual resources like timber, minerals, forage for livestock, and scenery) has virtually given way to the more systematic view of ecosystem management, managing the patterns and pro-cesses in a holistic manner to provide for sustained character and function, as well as for benefits and commodities for humans" (p. 256). EM underlies the decision in January 1998 by Forest Service Chief Mick Dombeck to place an 18-month morato-rium on road building in some roadless areas within National Forests.

Thomas (1997) maintains that "... ecosystem management is only a concept for dealing with larger spatial scales, longer time frames, and many more variables (ecological, economic, and social) than have commonly been considered in past management approaches" (p. xi). Successful EM attempts to manage for sustain-able productivity of the whole ecosystem (Schowalter et al. 1997).

What does EM portend for soil and water conservation? I believe that EM is compatible with soil and water conservation because soil and water resources are critical elements of ecosystems. This position is supported by examining the role of soil and water resources in the provision of the ecological services described earlier. First, air and water purification is inextricably linked to soil and water resources. Good soil is essential for plant growth and healthy plants have the capacity to purify air and water. Plants are also critical determinants of climate, which in turn impacts air and water resources.

Second, mitigation of floods and drought is significantly influenced by land use/ management in upland and bottomland areas and land use/management is directly affected by the distribution and use of soil and water resources. Conversion of bot-tom lands and floodplains to agricultural production and the damming of rivers to provide irrigation water, hydropower, navigation, water supply and flood control directly affect the frequency and severity of flooding and drought. Both have a negative impact on food and fiber production. Conversion of floodplains and wet-lands to agricultural production dramatically alters the soil and water resources in those areas. Deforestation in many countries has increased soil erosion and reduced soil productivity.

Third, generation and renewal of soil and the management of soil resources jointly determines soil productivity. When rates of soil erosion exceed rates of regeneration, topsoil is depleted, which reduces the ultimate productivity of soil. Fourth, soil and water resources have an important role in detoxification and decom-position of wastes. Increasing attention is given to detoxify contaminated soil by using it to grow crops that can remove the toxins. Additionally, soil has the capac-ity to decompose wastes. Wetlands in particular, which contain hydric soils, have the capacity to remove nutrients, pesticides, and harmful bacteria from agricultural and urban runoff and from wastewater. Land with healthy soils and plants is widely used for the safe disposal of animal wastes. In fact, animal waste is a natural fertil-izer that adds nutrients removed by plants and other biophysical processes.

Fifth, maintenance of biodiversity of plants and animals is critically dependent on the quantity and quality of soil and water resources. Excessive soil erosion in managed landscapes and major changes in land use/management in an area can adversely impact endangered plant and animal species. Use of water resources, particularly in arid areas, can have a profound impact on biodiversity.

Sixth, soil and water resources have a vital role to play in partial stabilization of climate. While increased use of fossil fuels is the main culprit in the destabilization of climate, notably global warming, global climate treaties recognize the importance of vegetation (especially trees) as carbon sinks. Vegetation is determined by the distribution of soil and water resources. Global climate treaties allow net additions to forests to be credited toward prescribed reductions in carbon dioxide emissions. In this regard, deforestation has a de-stabilizing effect and re-forestation has a stabilizing effect on climate.

In summary, soil and water conservation/management directly affects our ability to achieve the goals of EM. There is a fundamental difference, however, between soil and water conservation/management and EM. We have over 50 years of experience in developing soil and water conservation practices and programs and in evaluating their adoption and effectiveness. Experience with EM is much more limited. Furthermore, failures appear to be more prevalent than successes. Based on an examination of 23 managed ecosystems, Holling (1995) concluded that management set the stage for ecological collapse. In particular, management resulted in less resilient ecosystems, more rigid management institutions and greater community dependence on natural resources. While the potential benefits of EM are significant, the challenges of successful implementation are considerable. Fortunately, our ability to meet this challenge is improving due to rapid advancements in scientific knowledge, information management and communication technologies. Successful EM will require developing a dynamic learning environment in which management is responsive to ecosystem change.

References

Daily, G.C. 1997. Introduction: What are ecosystem services? *In* Natures Services: Societal Dependence on Natural Ecosystems, ed. G.C. Daily, 1-10. Washington, DC: Island Press.

Diaz, N.M., and S. Bell. 1997. Landscape analysis and design. *In* Creating Forestry for the 21st Century: The Science of Ecosystem Management, ed. K.A. Kohm and J.F. Franklin, 255-270. Washington, DC: Island Press.

Haeuber, R., and J. Franklin. 1996. Forum: Perspectives on ecosystem management. Ecological Applications 6:692-693.

Holling, C.S. 1995. What barriers? What bridges? *In* Barriers and Bridges to the Renewal of Ecosystems and Institutions, ed. L.H. Gunderson, C.S. Holling, and S.S. Light, 3-25. New York: Columbia University Press.

MacKenzie, S.H. 1996. Integrated Resource Planning and Management: The Ecosystem Approach in the Great Lakes Basin. Washington, DC: Island Press.

Marsh, George P. 1865. Man and Nature. New York: Charles Scribner.

National Research Council. 1992. Restoration of Aquatic Ecosystems: Science, Technology, and Public Policy. Washington, DC: National Academy Press.

Schowalter, T., E. Hansen, R. Molina, and Y. Zhang. 1997. Integrating the ecological roles of phytophagous insects, plant pathogens, and mycorrihzae in managed forests. *In* Creating Forestry for the 21st Century: The Science of Ecosystem Management, ed. K.A. Kohm, and J.F. Franklin, 171-189. Washington, DC: Island Press.

Thomas, J.W. 1997. Foreword. *In* Creating Forestry for the 21st Century: The Science of Ecosystem Management, ed. K.A. Kohm and J.F. Franklin, ix-xii. Washington, DC: Island Press.

Williams, J.E., C.A. Wood, and Dombeck, M.P., ed. 1997. Watershed Restoration: Principles and Practices. Bethesda, MD: American Fisheries Society.

ORIGINAL PUBLICATION DETAILS

Tony Prato, Professor and Director, Center for Agricultural Resource and Environmental Systems, University of Missouri, Columbia, Missouri.

Prato, Tony. 1999. Soil and water conservation is essential for ecosystem management. *Journal of Soil and Water Conservation* 54(3):522-523.

Part 4
The Future of the Conservation
Land Ethic

Leopold as Practical Moralist and Pragmatic Policy Analyst

Bryan G. Norton

Editor's note. In this essay, a case is made that Aldo Leopold was an early proponent of adaptive ecosystem management. Viewing Leopold from this lens allows us to see, in the face of the current development of adaptive management and understanding of ecosystem services, that Leopold's vision is still not only relevant but largely yet unrealized.

Aldo Leopold's ideas and pronouncements on environmental policy, read fifty years after his death, establish how far Leopold was ahead of his—and our own—time. The ideas expressed in Leopold's writings draw upon his experience with a broad spectrum of conservation issues. They demonstrate how Leopold's approach, if followed, would immensely improve the process and substance of environmental policy, even today. Although Leopold was not a philosopher, he developed a remarkably complex and subtle "philosophy" of environmental management. He loved to speculate on "big"—or as he often said "general"—ideas, but he was much more than a prophet of a future environmental consciousness. The ideas he lived by were the ideas that were forced upon him by years of thoughtful and painful experience. His discussions of policy often read like briefings he might like to have given to his first boss, the eminently practical Gifford Pinchot. In these discussions Leopold generally eschewed "intangible" ideas, accepting common philosophical and religious commitments as constraints on his speculations; yet he gave—or struggled valiantly to give—carefully articulated reasons and justifications for all of his management precepts.

It may be helpful to list some of the ideas, articulated between 1920 and his death in 1948, that establish Leopold's claim to prescience in the area of management theory and process. First, he insisted—contrary to his contemporaries and in opposition to most of today's congressional representatives—that ethics, not economics, ultimately validate environmental policies. Second, in anticipation of the current trend toward public and stakeholder participation in policy process, Leopold expressed his progressivist-populist faith that it must be farmers, sportsmen, and other citizens themselves who accomplish conservation. This second belief led to a third idea, one that is anathema to many environmental managers today, as it was to Leopold's own contemporaries in government resource agencies. He believed that public-servant environmentalists should be just that; that the highest calling of resource managers was education and public involvement, rather than what he derisively called the "ciphers" of management economics (Leopold 1924b, pp. 126-127). Fourth, Leopold recognized before others that management cannot simply be scientific in the sense of *applying* fixed principles of science; rather, and more important, we should be *managing* scientifically in the dynamic sense. Leopold thus insisted on policies designed to get results and reduce our ignorance through experiments with real controls.

Similarly, today's still-nascent but increasingly important ideas of ecosystem management were given shape by Leopold in his relentless attacks on atomistic management, which separated management of the land into "many separate field forces" (Leopold 1933b). He advocated instead an integrated approach to the management of resources. To these innovations we could add mention of Leopold's extraordinary concerns for our resource legacy to future Americans and a coherent and reflective concept of "sustainable development."

Each of these ideas—and there are others—would have qualified Leopold as an important innovator. But the totality of them, and the way in which Leopold used his unparalleled powers of observation to illustrate, sharpen, and weave these points together, mark him as the premier genius in the field. We would do well to listen to him very carefully when we choose actions to alter or "improve" on nature and natural functioning of ecosystems.

I regard Leopold the policy analyst, the policy-maker, and the practical moralist as the originator and spiritual father of the flourishing tradition of "adaptive ecosystem management," so ably espoused today by C. S. Holling, Carl Walters, Kai Lee, and others (see Norton 1996). Scientifically, Leopold anticipated the idea of ecological resilience, so prominent in the writings of Holling and the adaptive managers, when he described semiarid countries "set in a hair-trigger equilibrium" (Leopold 1924a, p. 112; see Holling 1995). He clearly recognized that shortsighted management could render ecological systems and processes vulnerable to collapse. Leopold also anticipated a unifying theoretical idea, which later came to be called "general systems theory," or (in theoretical ecology) "hierarchy theory." Leopold's brilliant insight—that managers and agriculturalists must, to be successful, "think like a mountain"—was not (or at least was not only) a mystical vision (Leopold 1949, p. 132; see Norton 1990). It was hard-won wisdom, that: (a) the manager, who observes and manipulates, is a part of the system, and not only views it but changes it from within; and (b) we can understand observed nature more coherently if we see it as a nested hierarchy of subsystems, with larger, slower-moving systems forming the "environment" for the smaller systems that compose it.

This scalar analysis, which would be incorporated wholesale into adaptive management, was embodied in Leopold's brilliant simile. In "Thinking Like a Mountain" Leopold provided a case study. The policy of killing wolves to increase deer herds seemed, from a short-term, human perspective to be a good idea to Leopold and others. He learned, however, that good policy must pass not only the short-term test of human economic reasoning but also the test of the mountain, which requires an ecological and evolutionary perspective. In addition to anticipating the adaptive managers' ideas of multiscalar analysis, Leopold also foresaw their emphasis on the need to include citizens and stakeholders in an iterative process that, at its best, involves social learning and the development of locally effective and cooperative institutions.

Leopold was a "scientific" manager who eventually, and sometimes through painful experience, came to appreciate both the strengths and the weaknesses of science in the practice of management. Because he had the professional mentality of a manager before becoming a professor, Leopold developed a nose for relevant science. Management, for him, asked interesting questions of ecology, and ecology provided useful tools for potential application. Accordingly, Leopold was scornful of pure theory when unrelated to practice, complaining that we argue over our abstract and conflicting ideas as to what needs to be done; instead, he suggested, we should "go out and try them" (Leopold 1930). He enjoyed speculation, but was disdainful of it when it was cut loose from experience, and from pressing environmental problems.

Much has been written about Leopold's science and about his moral beliefs. Few authors and commentators, however, have acknowledged that Leopold expressed a quite sophisticated philosophy of science and epistemology, which was for him intertwined with his management philosophy. Because he functioned as a manager in both the political and scientific worlds simultaneously, he constantly faced both controversy and uncertainty. Consequently, he judged facts by their usefulness. In the 1923 manuscript "Some Fundamentals of Conservation in the Southwest" Leopold expressed a sophisticated version of Darwinian epistemology, using it to cut through uncertainty about broad theoretical principles. Ultimately, he reasoned, our scientific and managerial behavior must be adaptive or we will not survive as a culture; we will "be judged in 'the derisive silence of eternity'" (Leopold 1923, p. 97). However, his healthy respect for the uncertainties of management left him wary of general pronouncements that lacked tight connection to actual experimental test. His concluding discussion of conservation morality in "Some Fundamentals" includes no less than five cautionary statements about the limits of philosophical and speculative language.

Leopold implicitly practiced (even as it was being articulated by Herbert Simon and others) the decision method known as "satisficing'" (Simon 1945). In "Game and Wild Life Conservation" he concluded that the only way to protect any wildness is "to set up within the economic Juggernaut certain new cogs and wheels whereby the residual love of nature" may be fractionally protected (Leopold 1932, p. 166). For Leopold satisficing meant triangulating through waters made choppy by mindless devotion to "progress" and boosterism, while relying on nothing more than the experimental spirit of science, democratic involvement, and faith in the traditions of good sportsmanship and husbandry expressed as respect for self and land. Satisficing—even moralistic satisficing in the style of Leopold—recognizes that policy cannot be guided by grandiose prior plans or principles, but must seek, with an eye toward best alternatives and pilot projects, to gradually improve policies and keep track of what works.

Although Leopold was quite successful in articulating a practical *and integrated* philosophy of environmental policy, confusion may be avoided if one thinks of Leopold's theories of management as quite distinct from his environmental ethics. My point is not that Leopold's views on value and management were unconnected, but rather that he related them differently than most professional environmental ethicists do today. Leopold worked from observation toward theory, whereas most of today's environmental ethicists attempt to establish universal principles and then "apply" them to particular cases. Leopold's powers of observation were legendary. As illustrated in his first attempt to survey conservation ethics in "Some Fundamentals of Conservation," his discussions of morality were usually preceded by a careful empirical analysis of trends and problems in resource use. He spent the first two-thirds of that landmark paper explaining what he saw from horseback in the vast lands under his management as director of operations for the Forest Service in the Southwest. He concluded that simple observation revealed damage from human use and that the damage was having economic impact. Ethics, then, were layered onto Leopold's practical, day-to-day approach of experimentalism and his daily recognition of the importance of economic motives. Leopold waxes philosophic, then, in order to explain and make sense of the whole field of environmental management; but he takes this to include economic and other established values, as well as new and more speculative ideas that might guide us to a more complete understanding of values and policy.

In "Some Fundamentals of Conservation" Leopold reviews ethical and metaphysical ideas—including anthropocentrism, organicism (the view that the earth itself is a living organism), nonanthropocentrism, and the possibility that "God himself likes to hear birds sing and see flowers grow"—as possible moral bases for conservation (Leopold 1923, p. 96). But he treats these ideas not as a philosopher would, trying to ascertain their truth based on *a priori* reasoning. Instead, he "screens" ethical beliefs for their policy usefulness, weighing their interest, plausibility, and verifiability, as well as their political appeal. So Leopold, writing from the perspective of a policy-maker in 1923, was willing to speculate on many ideas (indeed, he did so with obvious delight), but he began and ended his speculation with an anthropocentric framework. In the end he emphasized that the nobility we humans claim requires, independently of any moral demands placed on us by nature itself, that we mend our ways and protect ecological communities as we develop our lands. Along the way he considered and specifically dismissed organicism *as a guide to management*, because most managers believe "this reason is too intangible to either accept or reject as a guide to human conduct." He also adopted an agnostic position on the anthropocentrism-nonanthropocentrism debate, resolving that he would "not dispute the point" (Leopold 1923, pp. 95, 96).

Now we are in a better position to understand the relationship Leopold saw between ethics and policy. Most of the decisions he faced could, he thought, be guided by economic criteria, provided he took a long enough view of economics.

He also knew, however, that there were other decisions that, if made on strictly individual, economic grounds, would irreversibly damage the things he loved most about the region—the trout streams, the wild vistas, as so forth. Leopold, most basically, believed in the "convergence hypothesis" (Norton 1991). He believed that human interests and the "interests" of the natural world converge, and that if we were to protect humans—recognizing the full range of human values as projected into the indefinite future—one would also protect the natural world as an ongoing, dynamic biotic community. This leaves the hypothesis of intrinsic value in nature open, allowing Leopold to act in the long-term interests of humans as at least an approximation of what would be "good for nature." It also allows him, in policy contexts, to appeal to either human-oriented or nature-oriented explanations and justifications for the management goals he espoused, broadening the political base for conservation. He often speculated about values beyond the usual human-oriented values that dominated management in his day, but he carefully avoided resting controversial management proposals on these ideas. Leopold tried, whenever possible, to base real decisions on careful observation and experiment rather than speculation. This dualism, and this partial disengagement of Leopold's ethical thought from his management philosophy, explains how Leopold can both enjoy speculation and also wax disdainful of abstract thought.

Leopold saw, at least by 1933, that general philosophical solutions and "isms"—"Socialism, Communism, Fascism" and "Technocracy"—would fail, because they would not, or could not, adjust "men and machines to land" (Leopold 1933a, p. 188). Leopold knew long before the "isms" failed that "husbandry of somebody else's land is a contradiction in terms" (Leopold 1942, p. 298). But he also knew—contra conservative privatizers of today—that many public interests must be protected on public lands. He appreciated that communities of users must, with the help of sympathetic agency managers acting as teachers, maintain control of those public lands, using their government to realize community-based values and goals that go far beyond economic ones.

References

Holling, C.S. 1995. What barriers? What bridges? *In* Barriers and Bridges to the Renewal of Ecosystems and Institutions., ed. L.H. Gunderson, C.S. Holling, and S.S. Light, 3-34. New York: Columbia University Press.

Leopold, A. 1923. Some fundamentals of conservation in the Southwest. *Reprinted in* The River of the Mother of God and Other Essays, ed. Susan L. Flader and J. Baird Callicott, 86-97. Madison, WI: University of Wisconsin Press, 1991.

Leopold, A. 1924. Pioneers and gullies. *Reprinted in* The River of the Mother of God and Other Essays, ed. Susan L. Flader and J. Baird Callicott, 106-113. Madison, WI: University of Wisconsin Press, 1991.

Leopold, A. 1924. The river of the mother of God. *Reprinted in* The River of the Mother of God and Other Essays, ed. Susan L. Flader and J. Baird Callicott, 123-127. Madison, WI: University of Wisconsin Press, 1991.

Leopold, A. 1930. The American game policy in a nutshell. *In* Transactions of the Seventeenth American Game Conference (December 1-2, 1930), 281-283.

Leopold, A. 1932. Game and wildlife conservation. *Reprinted in* The River of the Mother of God and Other Essays, ed. Susan L. Flader and J. Baird Callicott, 164-168. Madison, WI: University of Wisconsin Press, 1991.

Leopold, A. 1933a. The conservation ethic. *Reprinted in* The River of the Mother of God and Other Essays, ed. Susan L. Flader and J. Baird Callicott. Madison, WI: University of Wisconsin Press, 1991.

Leopold, A. 1933b. Weather proofing conservation. American Forests 39(1):10-11,48.

Leopold, A. 1942. Land-use and democracy. *Reprinted in* The River of the Mother of God and Other Essays, ed. Susan L. Flader and J. Baird Callicott, 295-300. Madison, WI: University of Wisconsin Press, 1991.

Leopold, A. 1949. Thinking like a mountain. *In* A Sand County Almanac, 129-132. New York: Oxford University Press.

Norton, Bryan G. 1990. Context and hierarchy in Aldo Leopold's theory of environmental management. Ecological Economics 2:119-127.

Norton, Bryan G. 1991. Toward Unity among Environmentalists. New York: Oxford University Press.

Norton, Bryan G. 1996. Integration or reduction: Two approaches to environmental values. *In* Environmental Pragmatism, ed. Andrew Light and Eric Katz, 105-138. London: Routledge Publishers.

Simon, Herbert. 1945. Administrative Behavior. New York: Free Press.

ORIGINAL PUBLICATION DETAILS

By Bryan G. Norton, Distinguished Professor, School of Public Policy, Georgia Institute of Technology Atlanta, Georgia.

Norton, Bryan G. 1999. Leopold as practical moralist and pragmatic policy analyst. *In* The Essential Aldo Leopold: Quotations and Commentaries, ed. Curt Meine and Richard L. Knight, 201-205. Madison, WI: University of Wisconsin Press.

Toward a New Land Ethic

Norm Berg

Editor's note. Norm Berg, charter member of the Soil and Water Conservation Society and former chief of the Soil Conservation Service, wrote this piece, after passage of the 1996 farm bill, for *Conservation Voices*, a short-lived sister publication of the *Journal of Soil and Water Conservation*.

How relevant is Aldo Leopold's land ethic in present America? How many users of land and water, producing food and fiber for customers here and abroad, practice stewardship? In today's world, where decisions are driven more and more by economic necessity and reliance on the market system, rather than government, to protect and conserve renewable natural resources, will a land ethic suffice? It took a disaster, the Dust Bowl of the 1930s, to awaken the American public to the magnitude of the soil erosion problem. That awareness, building on the earlier work of Hugh Hammond Bennett, helped create the soil conservation movement. Federal soil and water conservation programs caused an awesome array of conservation practices to be installed on the privately owned lands in every region of the nation. Land managers had, as Leopold hoped, begun to see land not as a commodity belonging to individuals but as a community to which we belong and which we should use with love and respect. "That land is a community is the basic concept of ecology, but that land is to be loved and respected is an extension of ethics."

Until the enactment of the Food Security Act of 1985, which included a conservation title, soil and water conservation on privately owned agricultural land was voluntary, with technical and financial assistance available from USDA and increasingly from state and local government. Erosion had been accelerated by the fencerow to fencerow planting of the '70s, but now the 1985 farm bill allowed successful bidders to voluntarily enroll highly erodible cropland in the Conservation Reserve Program (CRP).

Three provisions of the bill, however—Sodbuster, Swampbuster, and Conservation Compliance—drastically changed the meaning of voluntary stewardship of agricultural resources. To qualify for federal farm program benefits, a set of rules and regulations came into play. The least demanding was Sodbuster, asking that any highly erodible grass or tree land be converted to cropland using a conservation plan. The Conservation Compliance provision was more stringent, requiring a plan to be in place by 1990 and fully implemented by 1995. The Swampbuster feature was a total prohibition on converting wetlands to cropland. All provisions related to the linkage of policies for commodities and conservation for the first time in USDA history. Thus, a land ethic or stewardship attitude was reinforced by the threat of loss of certain federal funds and possible sanctions if violations were detected or plans not adequately implemented.

The claim by those who manage land and water resources for agricultural production to be the nation's first environmentalists is credible enough if we accept available National Resources Inventory data and reports. There are critics of Compliance, because of the limited penalties. However, crop land soil loss is reduced, especially on lands in the CRP. Conversion of private agricultural wetlands has been reversed to the point that the no-wetland loss objective of a prior administration has nearly been achieved. The threat of increased regulation to combat nonpoint source water pollution has not diminished, but regulation is not the first option. The 1990 farm bill strengthened the 1985 Conservation Title, adding the Wetland Reserve Program (WRP). There was much debate about nonpoint sources of water pollution, though little action because of lack of adequate funding.

The 1996 farm bill has an excellent conservation title that include several incentive-based provisions. The process of rule-making and implementation is now underway. It contains some encouraging provisions, such as the promise of CRP and WRP programs until 2002; but the Freedom to Farm bill, with the drawdown of stocks for corn and wheat together with present attractive market prices, may tempt some to again crop marginal soils with less than adequate conservation measures.

As Leopold stated in his land ethic, "A system based solely on economic self-interest is hopelessly lopsided. It tends to ignore and eventually to eliminate many elements in the land that lack commercial value, but that are (as far as we know) essential to its healthy functioning. It assumes, falsely, I think, that the economic part of the biotic clock will function without the uneconomic parts."

Given that we are now looking at potential significant reductions in federal agricultural and conservation funding, as well as several proposals to reduce regulation, will a land ethic henceforth be the primary motivation to protect the soil and water resources used or agricultural production?

What do we understand about stewardship?

The President's Council on Sustainable Development, after two years of research and public hearings, released their report in February 1996. This presidential advisory panel said, "To achieve sustainability, institutions and individuals must develop [a] new way of thinking." It was their conviction that economic, environmental, and social equity issues are inextricably linked and must be considered together. Individuals and institutions must both take on greater responsibility and embrace an ethic of environmental responsibility.

Chapter 5 of this report begins, "Stewardship is an essential concept that helps to define appropriate human interaction with the natural world. An ethic of stewardship builds on collaborative approaches; and incentives in such areas as agricultural resources management, sustainable forestry, fisheries restoration, and biodiversity conservation. America is blessed with an abundance of natural resources which provides both the foundation for its powerful and vibrant economy and serves as the source of aesthetic inspiration and spiritual sustenance for many. Continued prosperity depends on the country's ability to protect this natural heritage and learn to use it in ways that do not diminish it."

This is followed by a challenging observation:

"Stewardship is at the core of this obligation. Without personal and collective commitment, without an ethic based on the acceptance of responsibility, efforts to sustain natural resources protection and environmental quality cannot succeed and as the population increases, so too will demands for food and fiber, fertile soil, clean and abundant water, and wildlife."

The advisory panel's review of incentives for stewardship also examined the need to eliminate any disincentives for natural resources stewardship. We have been down that road before, beginning with the 1985 Food Security Act. Subsidies, expenditures, and taxation were identified as three distinct areas for reform to become more of an incentive to stewardship. Appropriate regulations were part of the answer.

Are you still seeking, as I am, a definition of a land ethic or stewardship that will be meaningful to a private land owner or user of land and water for agriculture?

A review of land-use history since our nation began is dominated by controversies and conflicts of interest.

In the beginning, the West, seeking new settlers and development, wanted low land prices. The East feared the loss of its population and increased political power in the West. Our land history is replete with needed action always late. What we have had have been trends in the prevailing attitude toward the land. Hugh Hammond Bennett awakened the nation to the ravages of water, wind, and neglect that were damaging our land resources. He not only saw the need, but also had the ability to bring together the programs that would ensure a good stewardship of the land. Does each of us view "an eternity in a grain of sand, infinity in a blade of grass"?

The time is ripe for a discussion on stewardship. As the world moves rapidly into the future with an additional 87 million added each year, will any ethic prevail? Food security issues worldwide will challenge stewardship of land and water. Ethics may be the glue that holds a civilized society together, but ethics are not laws: they are a set of moral principles and values that govern an individual's or group's conduct.

ORIGINAL PUBLICATION DETAILS

By Norm Berg, Soil and Water Conservation Society Liaison, Washington, DC.

Berg, Norm. 1998. Toward a new land ethic. *Conservation Voices* 1(3):3-4.

Of What Value Are Values in Resource Management?

Pete Nowak

Editor's note. This piece recommends a shift in emphasis to improving the means—for example, the scientific knowledge base for conservation targeting—of effectively achieving conservation values–based objectives. Be sure to read past the third paragraph.

V alues, widely accepted statements about what is right and important, are truly important in terms of our everyday activities in natural resource management. Relevant examples in natural resource management include stewardship and sustainability.

We all recognize the preeminence of these values. A dominant part of our role as professional resource managers must be to support and enhance these values among the various publics with which we work.

By supporting and enhancing these values I mean they must be a principal part of both our professional and personal relations. Meetings, discussions, newsletters, and demonstrations must be built around such values as stewardship and sustainability. We also must continue and expand the time and resources we commit to working with youth groups relative to these values. I can think of no higher mission for resource professionals than instilling these values among youth, our future land users.

Reality

If you are reading this and feeling comfortable while agreeing with these statements, then I am afraid that I must disappoint you. Everything I have written up to now is either factually wrong or is a statement I do not support.

The concept of values is not found in the Soil and Water Conservation Society's *Resource Conservation Glossary*, Black's *Agricultural Dictionary*, or in SWCS's recent special publications. Moreover, I feel we already waste too much time and effort preaching values among the true believers and the converted. The same can be said of the immensely popular activity of working with youth groups. Proportionately too much time is spent working with these easy, captive, young audiences instead of focusing on the individuals who are actively abusing the land. Unlike youth groups, these abusers can be difficult, obstinate, and slam the door in your face while telling you to go to hell.

In sum, too many people spend too much time preaching and sermonizing on environmental values while avoiding the complexity associated with understanding why behavioral change does not follow. While the claim of the moral "high ground" on the basis of values may be personally gratifying, it has diverted needed attention away from understanding why land users may be unwilling or unable to act on these values.

I am questioning the very role of values in natural resource management. I do not blindly accept the importance or utility of dominant values, such as stewardship or sustainability. I should qualify this critical stance by noting that I am not an expert or even a student in the study of values. Values were not part of my academic program of study nor are they the focus or even part of my research today. I study statistical and spatial patterns of agronomic behavior within specific physical and institutional settings. Although I hear values frequently mentioned, I wonder about and even question their applicability to my daily duties and responsibilities. It is on the basis of this rather extended preface that I am now ready to address my subject: of what value are values in natural resource management?

Two Levels of Values

I believe, again remember that this is based on the assessment of a practitioner not an expert, that values operate at two levels in the natural resource management arena.

First, values are used to express the objectives to which we strive. Examples of these would be stewardship and sustainability.

Second, values can also be used to guide the process by which we attempt to achieve these objectives. While a significant amount of attention has been placed on values to represent objectives, consideration is rarely given to the values of describing how we achieve these higher objectives. For example, we have all heard, read, and seen countless appeals to stewardship and sustainability, however these objectives may be defined. Yet when these spokespersons advance stewardship and sustainability, how often do you hear an accompanying analysis on the values needed to achieve these objectives? Rarely, as it appears the values associated with policy objectives and the values needed for program implementation are largely divorced from each other.

What do I mean by values guiding program implementation procedures? At minimum we should consider the values of equity, efficiency, effectiveness, and that of knowledge itself. We cannot realistically expect to achieve sustainability or widespread stewardship without also incorporating widely held beliefs about equity, efficiency, effectiveness, and a sound scientific base of knowledge. Each of these deserves fuller explanation.

Equity

Equity is often interpreted as some form of equality. Yet equality, in the true sense of the word, is not possible in natural resource management. If one considers the tremendous diversity of the natural resource base (soils, landscape, hydrology, biotic communities, and weather attributes) and superimposes on top of that the tremendous diversity of natural resource users and their organizations (socioeconomic, firm, institutional, and cultural characteristics), then the concept of a true equality is nonsense.

Instead of equality we need to talk about an equitable access to stewardship and sustainability. That is, our programs need to be impartially targeted toward those with the greatest need, however that may be defined, not just those who have been past cooperators and show up at our offices and meetings. Nor should we target those with a larger than average number of acres or some other scale of operations indicator. And this last statement is in direct contradiction to the policy of many natural resource and educational agencies that still reward employees on the number of acres protected or changed, not which acres were protected or changed.

Equity in access to the opportunity to achieve stewardship or sustainability is based on the very simple notion that within any jurisdictional area we know which acres have the greatest need for resource management practices—not land classifications, such as highly erodible land, but land management units as recognized by the land user. For example, the land operated by Farmer Smith needs to be addressed first, followed by the land operated by Farmer Brown, and so on. The fact that this is not possible in most areas of natural resource management tells me that we have paid little attention to the value of equity in past deliberations. We have been too busy preaching stewardship to those who will listen, not those who need it.

Efficiency

The second value I put forward is that of efficiency—not an engineering efficiency represented by a mathematical ratio of input and output amounts, but an efficiency based on ecological, social, and economic parameters. Therefore, this will not be an efficiency represented by a cost-benefit analysis on the amount of dollars spent to protect "X" tons of soil or to lower the nitrate-nitrogen contamination in drinking water "X" parts per million. The value of efficiency I am referring to is one where a comprehensive understanding of what induces behavioral change is used to guide the remedial programs.

This remedial program, unlike past and current examples, will not "shotgun" technical, financial, educational, and regulatory inducements at generalized audiences or even priority areas. Instead, using assessment techniques, it will direct the appropriate mix of tools to meet the specific needs of individual land users operating in specific physical settings.

If one thinks of the various program tools, such as technical and financial assistance, education, and regulation, as inputs and changes in land user behavior as outputs, then efficiency refers to using different combinations of tools with different land users to bring about the greatest amount of behavioral change. This is efficiency in natural resource management.

While I believe that we have the scientific base to accomplish this, what has been lacking is the widespread belief in the value of efficiency needed to initiate this process. Instead of using an interdisciplinary scientific basis undergirded by the concept of efficiency, it appears we have been content to throw dollars, FTEs, and regulatory threats at problems and then bemoan the fact that farmers do not practice stewardship or adhere to the principles of sustainability.

Effectiveness

Effectiveness is different than efficiency in that rather than focusing on the effort required, it is based on the extent or degree to which we achieve goals or objectives. How effective are the conservation provisions of the farm bills or the institutional arrangements in various states and provinces at achieving stewardship or sustainability? Do we achieve stewardship or sustainability by focusing on saving soil or protecting water quality? I think not.

Higher values, such as stewardship or sustainability, are achieved by inducing long-term changes in human behavior. One-time estimates of tons of soil saved are poor indicators of the nature, extent, or permanence of change in land user behavior. By focusing on physical parameters as measures of program effectiveness, we are conveniently ignoring the potential yet probable concurrent changes in markets, technology, weather, institutional settings, and shifts in land ownership patterns. All these larger processes influence land user behavior, which, in turn, has impact on various physical parameters.

Evaluating program effectiveness based on physical parameters is a deficient concept. If this were not the case, then why do we constantly have to create new programs to address old problems that largely were solved at some time in the past? I would argue that it is because we have placed a low value on effectiveness conceived as an indicator of long-term change in human behavior.

We simply cannot separate the land user and the resource base he or she influences when discussing effectiveness in natural resource management. Even the most famous proponents of stewardship and sustainability recognized that the social, physical, and biological components of these concepts are intricately intertwined.

Why, then, does the organization of our sciences and the analytical framework of our resource programs act as a "meatgrinder" that rips and shreds these relations apart? As crude as it may sound, and hopefully this does not come as a surprise, resource managers are not in the business of making sausage. We are here to design and enable programs that can help land users achieve stewardship or sustainability. That is not going to happen with a myopic and reductionist perspective that separates the resource users from the resources they manage.

The effectiveness of any natural resources program must be determined by examining the changing relations between the land and the land user. Focusing on either one at the exclusion of the other produces meaningless results.

Knowledge

The last value that deserves more attention in our discussion of natural resource values is that of knowledge. Stewardship or sustainability need to be developed around a sound scientific foundation.

A sustainable resource base will remain ambiguous at best if it continues to be developed around political maneuvering, segregated interests, and emotional appeals. The value of sustainability or stewardship will be advanced when we can begin to answer questions as to whether a specific farm or ranch behavior within a specific physical setting adds to or subtracts from sustainability or stewardship. What I am asking is whether we have the scientific knowledge base to define the boundaries of sustainability or stewardship? Again, I think not.

What of Values?

So, of what value are values in natural resource management? Based on this brief assessment, I must honestly answer, very little.

While values, such as stewardship and sustainability, are valuable objectives, they will remain largely meaningless until we begin to pay more attention to the complementary values that will focus on the means to achieving these higher goals.

I do not want to be subjected to another layperson sermon on the moral imperative associated with stewardship, or for someone to use the latest "buzzword" of sustainability without spending an equal amount of time telling me how to achieve them. I have seen little change in the focus or method of research by my colleagues in the last few years, except that the term sustainable is now in the title. Fuzzy and abstract notions, such as sustainability, do not tell me how to work with a farmer, where to orient my future research, or help me work with an agency that tries to be all things to all people while doing none of it well. Value-driven objectives only become relevant when accompanied by additional values specifying how these larger goals can be realized.

Farmers, implicitly or explicitly, already work with the values of equity, efficiency, effectiveness, and knowledge. They attempt to differentially apply inputs to the land that has the greatest needs and capabilities (equity); all farmers recognize their best or most efficient fields or livestock; they know if they have been effective when it comes time to pay bills; and they place a high value on knowledge, both experiential and more formal recommendations coming from the private and public sectors.

Consequently, I do not feel it is asking too much to have our various resource management agencies adopt and evaluate the extent they are also equitable, efficient, effective, and knowledge-based.

Finally, I do not think it is possible to achieve sustainability or widespread stewardship by preaching, sermonizing, educating, or propagandizing farmers into adopting a land ethic. I say this because any farmer also has something that may be called a market ethic, a family ethic, a community ethic, and so on. Not only does the typical land user hold multiple values or ethics, that land user also recognizes that these values can often conflict with or contradict each other.

The values of stewardship or sustainability evidently come out fairly low when compared to these other values. For example, one can only wonder about the wisdom of a national agency that pays farmers to maintain base acres in cash grains at the same time it funds an independent sustainable agriculture program. Farmers live with value conflicts such as these on a day-to-day basis and respond accordingly. Instilling or strengthening stewardship or sustainability values does not resolve these underlying value conflicts.

Values are best left in academia. I say this because believing in a value and being able to act on that value are two separate processes. We have been either unable or unwilling to recognize this basic fact.

Continued boosterism of values, such as stewardship and sustainability, only serves to obscure contradictions in market and resource policies, to rationalize rhetoric rather than promoting reason, and to justify mediocrity in organizations with natural resource responsibilities. I will begin to pay more attention to such concepts as stewardship and sustainability when agencies, administrators, and agricultural policies begin to pay more attention to the values of equity, efficiency, effectiveness, and knowledge. I look forward to that day.

ORIGINAL PUBLICATION DETAILS

Pete Nowak, Professor, Department of Rural Sociology, College of Agricultural and Life Sciences, University of Wisconsin, Madison, Wisconsin.

Nowak, Pete. 1992. Of what value are values in resource management? *Journal of Soil and Water Conservation* 47(5):356-359

Science and Stewardship in a Nonmonolithic Conservation Movement: Facilitating Positive Change

Mark Anderson-Wilk

Conservation has a long history of bringing together diverse interests around a common cause. But "the conservation movement" is, of course, neither singular in what is to be conserved nor in the philosophical principles underpinning it.

Here I attempt to explore some of the variations in conservation ethics that have made the conservation movement complex, compelling, and challenging to those involved.

United and Divided in Conservation

One of the long-existing philosophical tensions within the conservation movement has been between a preservationist strain (with the objective of protecting resources, habitats, and species for their own sake) and a "use but use sustainably" strain (with the objective of conserving natural resources so that humans can continue to use them productively into the future).

Such philosophical contrasts existed over a century ago as personified by John Muir and Gifford Pinchot. A number of environmental historians and philosophers (e.g., Miller 1991; Norton 1991; Katz 1996) have described the friendship and philosophical parting of ways between Muir and Pinchot. Muir founded the Sierra Club and is well known for his writings on wilderness preservation and for advocating the establishment of national parks. Pinchot was the first chief of the US Forest Service, a founder of the Society of American Foresters, a friend and advisor of Theodore Roosevelt, and a vocal conservation advocate.

While Muir advocated for the preservation of pristine wilderness areas, Pinchot believed that conservation "recognizes fully the right of the present generation to use what it needs and all it needs of the natural resources now available, but it recognizes equally our obligation so to use what we need that our descendants shall not be deprived of what they need" (Pinchot 1910).

It has been suggested that the ethical divide between Muir and Pinchot was bridged by Aldo Leopold's land ethic (Norton 1991; Katz 1996). According to Katz (1996), since Leopold "includes in his calculations of ethical behavior the good for the nonhuman world (its 'integrity, stability, and beauty'), he can acknowledge the

116

kinds of goods that Muir sought to maximize (the beauty, harmony, and divinity of the wilderness) with the kinds of goods that Pinchot sought (the overall maintenance of the natural system as a reserve for resources for the future)."

Variations of the two conflicting conservation worldviews, however, have continued on.

Gould (1993) cautioned against versions of the land ethic focused too far into the future: "Environmental movements cannot prevail until they convince people that clean air and water, solar power, recycling, and reforestation are best solutions (as they are) for human needs at human scales—and not for impossibly distant planetary futures."

Wilson (2002) describes the need for a land ethic that combines short-term and long-term values: "To select values for the near future of one's own tribe or country is relatively easy. To select values for the distant future of the whole planet is relatively easy—in theory at least. To combine the two visions to create a universal ethic is, on the other hand, very difficult. But combine them we must, because a universal environmental ethic is the only guide by which humanity and the rest of life can be safely conducted through the bottleneck into which our species foolishly blundered."

Another example of the persisting plurality of views of conservation is the differing opinions on the relationship between economic factors and environmental stewardship. Dunlop (2006) argues that "if you have economic growth, you can be pretty sure you are going to have environmental improvement," while Daily et al. (1997) assert that "economic development that destroys habitats and impairs services can create costs to humanity over the long term that may greatly exceed the short-term economic benefits of the development."

Over the years, differing conceptions of conservation have become built into many institutional structures to fit their purposes and activities. For example, the US Department of the Interior, with its public land focus, and the US Department of Agriculture, with its private land focus, both have claims to conservation. The US Environmental Protection Agency also has a role in conservation, though the name of the agency contrasts with Pinchot's vision since it clearly indicates an intention to protect the environment from people, rather than future use from current use.

Conservation has been taught in a wide variety of university departments in part because conservation is a set of ethical principles that can be applied to a variety of areas of study. Shifts in the society's collective values and perspective on the environment are reflected in the changing of university department names over the years—many departments of forestry, for example, are now called departments of natural resources, which is not only broader but also changes the emphasis from the human use to the environmental resource.

The objectives and character of conservation policies and programs have also experienced gradual changes reflective of changing conceptions of conservation. Cox (2006) explains that, "for decades, the primary purpose of agricultural conservation programs was to improve the productivity of US farms and ranches and to protect the 'natural resource base' that sustained the agricultural enterprise. In the past 15 years, the environmental agenda—water quality, air quality, biodiversity, among other objectives—has superseded traditional objectives as the primary purpose of agricultural conservation programs."

Another example of this shift in society's conception of conservation is reflected in how words such as "green" are used. The "Green Revolution" of the 1960s and 1970s was focused on the development and distribution of robust crop varieties to help address hunger around the world. The "green revolution" today speaks to the public's desire for environmental protection and sustainability.

Current and near-future worldwide challenges related to the economy, energy, and climate change will likely exacerbate the shifts and variability of how conservationists and society at large view conservation. "As natural capital and ecosystem services become more stressed and more 'scarce' in the future, we can only expect their value to increase" (Costanza et al. 1997), and increasing competition for increasingly valuable resources will certainly have an impact on the role of conservation in the years to come.

The Scientist as a Human Being

Science slowly builds knowledge by methodically testing and modifying hypotheses. Good scientists use scientific methods, not value judgments, to collect and analyze data and are reluctant to extend their conclusions beyond the defined scope of their study. Many of the qualities that make good science good are quite different than the qualities that make advocacy and policy development effective. However, advocacy gains credibility through reliance on science, and science takes place in a context that involves value-based decisions: Which research priorities are funded is influenced by a complexity of interests, and the results of scientific research are used, often out of the hands of the scientists, to shape products, programs, and policies.

Noss (2006) notes that "one of the most hotly contested issues in ecology and conservation biology is about whether, or to what degree, scientists should be involved as advocates in public policy debates. Although the conventional wisdom has been to maintain a 'healthy distance' between science and policy, this view is increasingly challenged."

Many scientists are understandably reluctant to express their values or ethics (anything that might be perceived as a shade of ideology), with the exception of professional ethics such as the importance of treating human study subjects according to ethical protocols, acknowledging conflicts of interest, and preserving the integrity of peer review.

Scientists generally understand their role is to "investigate the way the world is but abjure as [scientists] any claim about the way the world ought to be" (Sagoff 2007). In other words, science does not and cannot tell us that we should protect the environment or that conserving natural resources is better or worse than depleting them. The decision to conserve must be made outside science, and then science can be used as an input to finding out the most effective way to conserve.

Koertge (2000) argues that "when political considerations are used to limit the questions that can be raised, the hypotheses that can be tested, or the alternative explanations that can be brought forward, that area of inquiry ceases to have scientific value, regardless of whether the political motivations are good ones or bad ones."

Some scientists may fear that if they expose their personal beliefs and sense of ethics, others might lower their perception of their scientific credibility. In the environmental sciences, in particular, some scientists may actually wish to disassociate with the environmental "cause" for fear of being perceived to be linked to environmental extremists whose advocacy is not always well supported by the scientific literature.

Nevertheless, "in recent years there have been increasing pleas for scientists to play a more active role in environmental policy discussions" (Minteer and Collins 2005).

With increasing pressures for applied research over pure research, a lot of science is being conducted not just for the sake of knowledge but to facilitate change. The scientist is thus already an instrument in a process and, some would say, can have additional usefulness in making a positive change happen more fully or more efficiently by expanding his/her role to include that of advocate or at least "recommender" (Shrader-Frechette 1996; Noss 2007).

Many of the areas of human concern that seek scientific input cannot practically wait to find out all that science will discover. Instead, a series of incremental changes are made based on the information available at any given time. Somebody has to extrapolate recommendations from a set of studies to the applied field. According to Noss (2006), "people who understand the science that went into generating facts are in the best position to interpret those facts and to recommend how they are applied to policy or management decisions."

The activity of the scientist outside the proper bounds of science may be justified by the person acting as a scientist while conducting science and acting as an individual while making ethical judgments or advocacy.

Gould (1996) provides the following example of a scientist not only being justified in "speaking up" as an individual related to an ethical issue but being duty-bound to do so: "As a scientist, I can refute the stated genetic rationale for Nazi evil and nonsense. But when I stand against Nazi policy, I must do so as everyman—as a human being. For I win my right to engage moral issues by my membership in Homo sapiens—a right vested in absolutely every human being who has ever graced this earth, and a responsibility for all who are able."

Sagoff (2007) suggests that this principle can be extended to scientists expressing their personal sense of values regarding the environment because "as feeling and emotional beings they recognize compelling aesthetic qualities in nature as well as their moral responsibility and free will as human beings."

For the sake of argument, however unrealistic, what if science were being used by a malevolent entity to investigate "best practices" for reducing natural resources and contaminating waters? Would the scientists involved be justified in speaking out against how the science was being used? On a similar note, if an entity is using science to protect natural resources, why shouldn't the scientists involved be justified in speaking up that they, as human beings, think the research findings are valuable and should be applied?

Those scientists working on conservation topics who don't feel comfortable acting as vocal advocates can still have a personal sense of purpose, satisfaction, and pride not only in doing good science but also that their science is being used to advance change in ways that support their own personal ethics.

Challenges to the Conservationists' Land Ethic

"Ethics enters the picture when we begin to think about the meaning and value of the use of the environment, when humans examine the nature of their relationship with nature" (Katz 1996).

In the conservation arena, the concept of land ethic or land stewardship has been important, among other things, as a motivator for land managers to adopt practices that are environmentally beneficial.

An ethic for the land has of course existed in many forms in many cultures throughout history, but the land ethic in America is usually traced through key philosophers in the conservation movement.

Robert Marshall, who was chief of the US Forest Service Division of Recreation and Lands, developed a concept of a land ethic in *The People's Forests* first published in 1933: "The time has come when we must discard the unsocial view that our woods are the lumbermen's and substitute the broader ideal that every acre of woodland in the country is rightly a part of the people's forests" (Marshall 2002).

The chief philosopher and scribe of the conservation land ethic was Aldo Leopold, himself a scientist. Leopold's land ethic "reflects the existence of an ecological conscience, and this in turn reflects a conviction of individual responsibility for the health of the land. Health is the capacity of the land for self-renewal. Conservation is our effort to understand and preserve this capacity" (Leopold 1949). Leopold (1949) explains that "the land ethic simply enlarges the boundaries of the community to include soils, waters, plants, and animals, or collectively: the land."

Leopold's son, A. Carl Leopold, himself a plant research scientist, states that "conservation has traditionally focused ... on reducing consumption and acting to preserve ecological remnants. Aldo Leopold's example added a distinctive new component to conservation: positive actions to restore ecological communities" (Leopold 2004).

The land ethic has inspired the passion of many scientists over the years, including Edward O. Wilson, the Harvard entomologist: "A global land ethic is urgently needed. Not just any land ethic that might happen to enjoy agreeable sentiment, but one based on the best understanding of ourselves and the world around us that science and technology can provide.... We will be wise to listen carefully to the heart, then act with rational intention and all the tools we can gather and bring to bear" (Wilson 2002).

Variations on Leopold's land ethic have been held up and echoed for decades. The land ethic has also been critiqued from several fronts.

For example, J. Baird Callicott (1983), a Leopold scholar and environmental philosopher, describes it as "yet another set of rules or limitations. It calls for obligation, self-sacrifice, and restraint and thus could be unappealing to farmers and landowners."

Callicott instead recommends focusing on Leopold's concept of a land aesthetic. He explains that the land aesthetic "may be more palatable because it emphasizes assets and rewards" and encourages landowners to have pride about contributing to life and beauty on their land (Callicott 1983). "The land aesthetic calls attention to the psychic-spiritual *rewards* of maintaining the biological integrity and diversity of the rural landscape" (Callicott 1983).

Others have argued that environmental ethics have "inspirational power, but they are more useful in preaching to the converted than in resolving controversy" or providing useful input to policy development (Shrader-Frechette 1995).

Stephen Jay Gould, the late Harvard paleontologist and natural history essayist, challenged the ethical concept of stewardship commonly used by conservationists: "We are one among millions of species, stewards of nothing. By what argument could we, arising just a geological microsecond ago, become responsible for the affairs of a world 4.5 billion years old, teeming with life that has been evolving and diversifying for at least three-quarters of this immense span. Nature does not exist for us, had no idea we were coming, and doesn't give a damn about us" (Gould 1993). (Note that Gould also spoke up against inappropriately using the point he makes here to excuse practices that are bad for the environment.)

Should the conservation community still be seeking new definitions of a land ethic and land stewardship, as Norm Berg (1998) was a decade ago? Is the land ethic "still the crucial moral counterforce in opposition to short-term expediency in determining the use of our land, air, and water resources," as suggested by Bruce Babbitt (1995)?

Can Value be Placed on Values?

Economic values and ethical values are, of course, very different things, but "the two are not necessarily unrelated" (Krupnick and Siikamäki 2007). Can ethical values be entered into consideration in the process of trying to quantify the economic value of environmental resources that are currently outside market transactions?

The concept of ecosystem services has become central to an ongoing quest to assign economic value to the natural environment.

Bingham et al. (1995) described two categories of ecosystem attributes—those that are *directly* useful (such as food, medicine, and recreation) and those that provide *indirect* benefits by helping to sustain viable ecosystems, which in turn translates into goods and services. Ten years later, the Millennium Ecosystem Assessment (2005) refined the system of categorizing ecosystem services into "*provisioning services* such as food, water, timber, and fiber; *regulating services* that affect climate, floods, disease, wastes, and water quality; *cultural services* that provide recreational, aesthetic, and spiritual benefits; and *supporting services* such as soil formation, photosynthesis, and nutrient cycling."

Costanza et al. (1997) highlighted the importance of understanding the value of ecosystem services: "Because ecosystem services are not fully 'captured' in commercial markets or adequately quantified in terms comparable with economic services and manufactured capital, they are often given too little weight in policy decisions."

Daily (1997) noted that "the importance of ecosystem services is typically quickly appreciated, but the actual assigning of value to ecosystem services may arouse great suspicion, and for good reason. Valuation involves resolving fundamental philosophical issues (such as the underlying bases for value), the establishment of context, and the defining of objectives and preferences, all of which are inherently subjective. Even after doing this, one is faced with formidable technical difficulties with interpreting information about the world and transforming it into a quantitative measure of value."

Significant work has been done in the past decade to attempt to better measure and quantify the value of ecosystem services (and to better account for ecosystem services with indirect benefits), but there is no promise of absolute scientific precision in any of the methods advanced.

Many difficulties exist in the task. For example, "different stakeholder groups often perceive very different costs and benefits from ecosystems" (Pagiola et al. 2004).

In addition, even the most comprehensive method may not adequately account for some factors that don't easily translate into economic values. This would then lead to the "question of whether an action (for example, policy protecting old-growth forest) is to be judged on its intrinsic rightness or based on the measurable benefits and costs that might result" (Hackett 1998).

Pagiola et al. (2004) cautions against reducing conservation decisions to economics: "Economic analysis is not and should not be the only input into conservation decisions. People can and do decide to conserve things based on a range of other criteria."

These cautions echo Leopold's statement that we should not assume "that the economic part of the biotic clock will function without the uneconomic parts" (Leopold 1949).

Many of those involved in assigning economic values to ecosystem services would acknowledge the imperfections and shortcomings, but some quantification is necessary as a practical basis for potential actions to be weighed and recommendations made.

Debating the Ethical Center

Economic ecosystem valuation methods are by and large anthropocentric (human centered) in that they assess the value of ecosystem services to humans either directly or indirectly.

Environmental philosophers distinguish between *instrumental value* and *intrinsic value*. Something has instrumental value if it provides something that is valuable (corn is valuable because it can be used for food and fuel). Something has intrinsic value if it has value in itself. The predominant culture is anthropocentric in that it generally assigns intrinsic value to humans alone and instrumental value to things only when they provide something that is valuable to humans.

Callicott (1989) describes an ecocentric movement that has "advocated a shift in the locus of intrinsic value from individuals (whether individual human beings or individual higher 'lower animals') to terrestrial nature—the ecosystem—as a whole."

Another of the many divides in the environmental movement has been characterized by a *shallow ecology movement* concerned with pollution and resource depletion's effect on humans and a *deep, long-range ecology movement* that has a "deeper" commitment to the environment that is not human centered (Naess 1973; Devall 2001). Deep, long-range ecology is based on an ecocentric ethic that living things "have inherent worth and we humans have duties to protect them" (Shrader-Frechette 1998).

While Aldo Leopold recognized the anthropocentric necessities of land management, his land ethic clearly called for a shift toward an ecocentric or biocentric ethic: "Obligations have no meaning without conscience, and the problem we face is the extension of the social conscience from people to land" (Leopold 1949).

Like the deep, long-range ecology perspective, Leopold sought to embrace the land as part of the community included in ethical considerations. The deep ecology concept "rejects the human-centered perspective on nature and shifts the ethical issue to a biocentric base. This point of view resonates with Aldo Leopold's ... land ethic" (Leopold 2004).

Society's pendulum continues to swing back and forth between anthropocentric and ecocentric values.

In 2007, Kareiva and Marvier reported that "a growing number of conservationists argue that old ways of prioritizing conservation activities should be largely scrapped in favor of an approach that emphasizes saving ecosystems that have value to people." Kareiva and Marvier frame their argument as a new approach to replace an old approach, although, as we've seen, these two approaches in one form or another have been struggling to win the minds and actions of conservationists since the beginning of the conservation movement.

Not so long ago, conservation was criticized for being too anthropocentric; now, it is said to have a "misanthropic reputation" (Kareiva and Marvier 2007).

Basing economic valuation methods on anthropocentric ethics rather than ecocentric ethics may have some practical advantages, but even an anthropocentric approach poses serious challenges. For example, "though few people have seen bald eagles or manatees up close, many want to make sure their habitats are preserved. Such non-use values are challenging to measure because they are not captured in market transactions or other observable choices" (Krupnick and Siikamäki 2007).

Kareiva and Marvier (2007) base their support for an anthropocentric ethics approach on what they see as the likely future conservation will face: "Because our environment will consist mainly of human-influenced systems, biodiversity protection must be pursued in the context of landscapes that include urban centers, intensive agriculture, and managed forests and rivers, not just nature preserves."

On a more philosophical level, McShane (2007) continues to uphold support for ecocentric ethics. He explains that the concept of intrinsic value can help us understand "the different ways that we do and should care" about the natural environment.

Cooperation and Conflict in Conservation

What about the potential for putting all these differences aside to work for the common cause of conservation? "Cooperative conservation" is currently a popular concept, and many forms of cooperation, including across public and private sectors, clearly offer a multitude of strategic, common sense benefits. Consensus models have gained popularity because they promise "win-win" outcomes (Peterson et al. 2005).

Some, however, argue that cooperation can too easily lead to compromise, complacency, and the "watering down" of environmental ideals. Devall (2001), for example, suggests that some types of cooperation may generate "mild reforms in public policy and practices [that] are basically useless."

Peterson et al. (2005) argue that resource management by consensus is dangerous because the attempt to placate everyone reduces any impetus for change and reinforces the status quo. Instead, they assert, "argumentation, and the science-based deliberation it entails, places conservation on more firm epistemological ground.... An emphasis on argument legitimizes and facilitates change, whereas an emphasis on consensus further legitimizes continuity or stability" (Peterson et al. 2005).

Conflicts and debates inside and outside of the conservation community may actually serve a function in advancing conservation.

While detraction, disagreement, and nuance continue today, there currently *appears* to be a public, corporate, and political consensus about the importance of the environment and the ethical responsibility of humans to minimize their impact on it.

The current general awareness of and enthusiasm for the environment is a good sign for conservation, but it has contributed to a loss of clarity in the language used to explain environmental concepts and commitment. For example, Willers (1994) describes that *sustainability* has become a "code for perpetual growth" used by global industrialists and land developers to make themselves look better to an ecocentric public. Similarly, mass numbers of organizations are self-identifying as "green," "eco-friendly," and "sustainable," yet for the majority of such claims, there is no scientifically based way to know how "green" one product or service is compared to another.

This fuzziness in terminology may be one small downside compared to the large upside that people, companies, and governments are pledging their support for the goals of conservation. A consumer "land ethic" has been sweeping across the nation and world. Environmental messages are flashing in many corners of the culture. Companies are now praised for being "environmentally friendly" and criticized if they are not. Businesses, schools, agencies, and individuals are making more and more choices with "green" on the label. If all this support and posturing really translates into a collective meaningful reduction in impact on the environment, then conservationists representing all camps of the movement have reason to be hopeful that a land ethic can drive conservation progress.

The key is to use ethics to set the right direction and science to make sure it is done right.

References

Babbitt, B. 1995. The land ethic: A guide for the world. *In* Aldo Leopold: The Man and His Legacy, ed. T. Tanner. Ankeny, IA: Soil and Water Conservation Society.

Berg, N. 1998. Toward a new land ethic. Conservation Voices 1(3):3-4.

Bingham, G., R. Bishop, M. Brody, D. Bromley, E. Clark, W. Cooper, R. Costanza, T. Hale, G. Hayden, S. Kellert, R. Norgaard, B. Norton, J. Payne, C. Russell, and G. Suter. 1995. Issues in ecosystem valuation: Improving information for decision making. Ecological Economics 14:73–90.

Callicott, J.B. 1983. Leopold's land aesthetic. Journal of Soil and Water Conservation 38(4):329-332.

Callicott, J.B. 1989. In Defense of the Land Ethic: Essays in Environmental Philosophy. Albany, NY: SUNY Press.

Costanza, R., R. d'Arge, R. de Greoot, S. Farver, M. Grasso, B. Hannon, K. Limburg, S. Naeem, R.V. O'Neil, J. Paruelo, R.G. Raskin, P. Sutton, and M van den Belt. 1997. The values of the world's ecosystem services and natural world capital. Nature 387:253:260.

Cox, C. 2006. Foreword. *In* Environmental Benefits of Conservation on Cropland: The Status of Our Knowledge, ed. M. Schnepf and C. Cox. Ankeny, IA: Soil and Water Conservation Society.

Daily, G. 1997. Chapter 1. *In* Nature's Services: Societal Dependence on Natural Ecosystems, ed. G. Daily. Washington DC: Island Press.

Daily, G.C., S. Alexander, P.R. Ehrlich, L. Gouder, J. Lubchenco, P.A. Matson, H.A. Mooney, S. Postel, S.H. Schneider, D. Tillman, and G.M. Woodwell. 1997. Ecosystem services: Benefits supplied to human societies by natural ecosystems. Issues in Ecology 2:1-16.

Devall, B. 2001. The deep, long-range ecology movement. Ethics and the Environment 6(1):18-41.

Dunlop, B.N. 2006. Conservation ethics. Society 43(3):13-18.

Gould, S.J. 1993. The golden rule: A proper scale for our environmental crisis. *In* Eight Little Piggies: Reflections in Natural History. New York: Norton and Company.

Gould, S.J. 1996. The most unkindest cut of all. *In* Dinosaur in a Haystack: Reflections in Natural History. London: Jonathan Cape.

Kareiva, P., and M. Marvier. 2007. Conservation for the people. Scientific American 297(4):50-57.

Katz, E. 1996. The traditional ethics of natural resources management. *In* Nature as Subject: Human Obligation and Natural Community. Lanham, MD: Roman and Littlefield.

Koertge, N. 2000. Science, values, and the value of science. Philosophy of Science 67:S45-S57.

Krupniac, A., and J. Siikamäki. 2007. How people value what nature provides. Resources 165:14-16.

Leopold, A. 1949. A Sand County Almanac. New York: Oxford University Press.

Leopold, A.C. 2004. Living with the land ethic. BioScience 54(2):149-154.

Marshall, R. 2002. The People's Forests. Iowa City, IA: University of Iowa Press.

McShane, K. 2007. Why environmental ethics shouldn't give up on intrinsic value. Environmental Ethics 29:43-61.

Millennium Ecosystem Assessment. 2005. Ecosystems and Human Well-Being: Synthesis. Washington, DC: Island Press.

Miller, C. 1991. Gifford Pinchot and the Making of Modern Environmentalism. Washington, DC: Island Press.

Minteer, B.A., and J.P. Collins. 2005. Ecological ethics: Building a new tool kit for ecologists and biodiversity managers. Conservation Biology 19(6):1803-1812.

Naess, A. 1973. The shallow and the deep, long-range ecology movements: A summary. Inquiry 1:16.

Norton, B.G. 1991. Toward Unity Among Environmentalists. New York: Oxford University Press.

Noss, R.F. 2007. Values are a good thing in conservation biology. Conservation Biology 21(1):18-20.

Pagiola, S., K. von Ritter, and J. Bishop. 2004. How Much Is an Ecosystem Worth? Assessing the Economic Value of Conservation. Washington, DC: World Bank.

Peterson, M.N., M.J. Peterson, and T.R. Peterson. 2005. Conservation and the myth of consensus. Conservation Biology 19(3):762-767.

Pinchot, G. 1910. The Fight for Conservation. New York: Doubleday.

Sagoff, M. 2007. On the compatibility of a conservation ethic with biological science. Conservation Biology 21(2):337-345.

Shrader-Frechette, K. 1995. Practical ecology and foundations for environmental ethics. Journal of Philosophy 92(12):621-635.

Shrader-Frechette, K. 1996. Throwing out the bathwater of positivism, keeping the baby of objectivity: Relativism and advocacy in conservation biology. Conservation Biology 10(3):912-914.

Shrader-Frechette, K. 1998. The rights of natural objects. In Environmental Ethics, ed. K. Shrader-Frechette. Oxford, UK: Rowman and Littlefield.

Willers, B. 1994. Sustainable development: A new world deception. Conservation Biology 8(4):1146-1148.

Wilson, E.O. 2002. The Future of Life. New York: Vantage Books.

ORIGINAL PUBLICATION DETAILS

Mark Anderson-Wilk, Editor, *Journal of Soil and Water Conservation*, Ankeny, Iowa.

Anderson-Wilk, Mark. 2008. Science and stewardship in a nonmonolithic conservation movement: Facilitating positive change. *Journal of Soil and Water Conservation* 63(5):142A-146A

Leopold's Land Ethic, Ecosystem Health, and the Challenge of Affluenza

Y.S. Lo

While the work of Aldo Leopold—particularly his land ethics—has inspired some philosophers and activists and called down rebukes and critique from others, the widest international impact of his work is found in fields beyond philosophy and environmental activism. Leopold inaugurated the idea of ecosystem health in a seminal remark in 1941 when he commented that "although the art of land doctoring is being practiced with vigor, the science of land health is a job for the future." His remarks on the science of land health, along with his practical admonitions to develop sustainable and naturally inspired farming systems, have inspired many people, especially biologists, conservationists, and agriculturalists, to think in new ways about the challenges of land management and nature protection. By the 1970s a new discipline, ecotoxicology, had emerged, which seemed to fulfil some of Leopold's requirements for being a science of "land health." Two central insights of ecotoxicology are (1) that components of ecological systems can be damaged through the cycling and accumulation of toxic material within the systems and (2) that monitoring certain populations belonging to a system can provide early warning of risks to other populations within the same system. If insight 2 is true, then the measurement of certain parameters for the key populations might be just as useful, objective, and reliable as a guide to the health of systems as the measurement of heart rate, blood pressure, and aspects of body chemistry is for the health of an individual organism.

Despite his continuing influence, it is hard nowadays to defend Leopold's views on the relationship of naturalness, biodiversity, and land health. When he claimed that the science of land health was a job for the future, he clearly regarded health as associated with the *integrity, stability, and beauty* of the land. His land ethic states that "a thing is right when it tends to preserve the integrity, stability, and beauty of the biotic community. It is wrong when it tends otherwise" (Leopold 1949, pp. 224-225). It would be convenient if the pursuit of land health corresponded with maintaining stable and naturally diverse environments. But as Leopold's major contemporary exponent, J. Baird Callicott, has admitted, the concepts of biodiversity and ecosystem health come apart. Quoting research by Allen and Hoekstra (1992), Callicott has argued out that the loss of the American chestnut from forests in the southeastern United States did not compromise the health of these forest systems. However, that loss occurred in a context of changing biodiversity; hence, biodiversity can change while health remains unaffected (Callicott 1995, p. 357). In fact, the general association of health with diversity is challenged by well-known cases: For example, a plantation woodland may produce healthier timber than an old-growth forest. In the latter, it is the decaying of trees and the spread of fungi and disease organisms that open up places for explosions of life and diversity at many levels from microorganisms up to invading plant species.

As Angermeir and Karr (1994, p. 692) have pointed out, *biological integrity* can be defined as "the capacity to support and maintain a balanced, integrated, adaptive community having a species composition, diversity, and functional organization comparable to that of the natural habitat in the region." Such a sense of the notion of integrity may be not far from what Leopold had in mind. Many healthy and rela-tively natural communities do have just such integrity. On the other hand, it is also possible for there to be mixed natural and agricultural systems that are very healthy by any standard measure but in which biodiversity is not comparable at all to that which originally characterized the region. As Leopold pointed out, "few acres in North America have escaped impoverishment through human use" (Leopold 1939). Recent trends in restoration, conservation, and organic farming have attempted to reduce the worst effects of large monocultures and intensive farming systems. Yet reduction in biodiversity—in comparison to the pre-agricultural situation—is a widespread, if not universal, feature of agro-ecological systems, which largely aim at sustainability and are regarded as successful when they produce increases in biodiversity above the norms that intensive farming systems permit (MacDonald and Service 2007).

Health of a system, then, is not necessarily correlated with some earlier mea-sure of species composition and diversity, and so not a necessary accompaniment of integrity in Leopold's sense. Now, this sets a challenge, not only for Leopold, but also for contemporary sciences of the land, those aimed at maintaining system-level health while promoting diversity and some level of "naturalness." Just as the notions of health and diversity come apart, so also must the concepts of natural-ness and health. Leopold's land ethic was premised on respect for and cherishing of things "natural, wild, and free." But many animals in a state of nature enjoy relatively poor health, and so also do plants. If we value biodiversity, and if we value the natural interplay between species, the duel for survival between larger organisms and the microorganisms which infect them, then we may have to tolerate levels of disease within populations that are far higher than need be.

Putting the point another way, with some management, various plant and animal diseases could be reduced and hence the aggregate health of populations could be improved. In facing the issue of ethics and the management of national parks one theorist, Holmes Rolston, has argued that pain and suffering in the wild ought not be relieved if doing so would interrupt ecosystemic and evolutionary processes (see Rolston 1992). Evolution is driven, as Darwin originally pointed out, by the "struggle for existence," and Rolston's recommendations on park management are intended to let the struggle continue. Yet consider the parallel case for human communities. Thanks to the application of medical science and technology, many human populations have managed to overcome many of the natural limitations on human flourishing, and in so doing have changed the nature of human life and expectations. If we are lucky, born with robust constitutions and escape various environmental challenges, then we can enjoy natural good health. If we are less lucky, we may still—thanks to medical interventions—enjoy good health, but at the loss of "naturalness" understood as freedom from human technological manipulation and intervention.

The contemporary inspiration of Leopold's views on the land ethic, conservation, and sustainable agriculture is that we should look with a certain skepticism at human attempts to manage nature and replace natural systems of evolution and control with human substitutes. Think of the "land" in Leopold's sense as a loosely and feebly organized system, a kind of superorganism. In this sense, it is not an aggregate of individual plants, animals, waters, and soils, but is a unitary thing whose health consists in maintenance of its own natural capacities for self-regulation and continuing identity. To use a more recent notion, the land is a *resilient* system, capable of repairing damage by means of its own natural processes, if only these can be left to operate. The land, or ecosystem, is thus similar to the human or animal body itself through being a self-maintaining and self-regulating system. Like the land, the body is able to maintain itself in the face of environmental challenge and various perturbations by returning close to some earlier state after such disturbance. Health, both for the individual and for the system, involves this capacity for self-repair and self-renewal not merely the absence of disease and illness.

When it comes to public health and to agriculture, the short-term view often takes precedence. What is good for the population, it is often thought, is the improvement of individual health and thus the elimination of widespread diseases and deficiencies within a given population. Leopold's work, however, encourages us to take the longer term perspective, to "think like a mountain" as he put it. Leopold was far too practical to recommend that we follow nature in everything, or that nature always knows best. He recognized the importance of land management and the significance of conservative and sustainable farming practices. Yet throughout his work there are always cautionary reminders that our own wisdom is limited. As "plain members and citizens" of the larger biotic community, our conception of what is good for our communities and what is good for the land may be a blinkered and narrow one. The concepts of health, diversity, integrity, and naturalness that are central to Leopold's work are still ones that intrigue, confuse, and puzzle writers today—and his writings are a constant source of fresh thoughts and new ideas in the face of just such bewilderment. Leopold's land ethic, then, and his associated views on the science of land health, are not solutions to a determinate set of problems so much as a continuing source of provocation and challenge to professionals working in applied ecology and agriculture.

While his work on the science of land health has been influential in establishing the practical applications of ecological sciences, Leopold's impact on environmental ethics and philosophy has been of wide significance in the development of contemporary environmental thought. During much of the 20th century, theologians, naturalists, ecologists, and foresters living in the United States had regularly made value pronouncements in connection with life, wilderness, or "the land" in general. The Scot John Muir (founder early in the century of the Sierra Club and often called the "father of American conservation") advocated an esthetic concern and quasi-religious reverence for nature, its glory and majesty, while deploring the utilitarian and economic approach to farming and the land. When he wrote in his *Sand County Almanac*, "that land is a community is the basic concept of ecology, but that land is to be loved and respected is an extension of ethics," Leopold was following squarely in the footsteps of Muir. By contrast, his practical emphasis on conservation and the science of land health resonates with the work of the pioneering conservationist Gifford Pinchot. It was left to Leopold's later followers, most prominently the philosopher J. Baird Callicott to try to fuse Pinchot's practical and "wise use" approach to environmental management into productive synthesis with the legacy of Muir's reverentialism. In common with both Muir and Pinchot, though, Leopold was clear that any ethic of the environment should start from a rejection of purely economic thinking:

> [Q]uit thinking about decent land use as solely an economic problem. Examine each question in terms of what is ethically and aesthetically right, as well as what is economically expedient. A thing is right when it tends to preserve the integrity, stability, and beauty of the biotic community. It is wrong when it tends otherwise. (Leopold 1949, pp. 224-225)

Philosophical commentators on Leopold found two elements of interest in this: first, a rejection of ethical individualism in favor of a form of holism, one in which the needs and interests of individuals must sometimes be sacrificed in order to protect larger biological wholes—like populations or systems; second, a doctrine of the intrinsic value of natural things. As noted by many writers, one implication of the land ethic is that an individual member of the biotic community ought to be sacrificed whenever necessary for the protection of the good of the whole community. For instance, Callicott once argued that if culling a white-tailed deer is necessary for the protection of the holistic biotic good, then it is a land-ethical requirement to do so. Once applied to the human domain, the thought that the individual should be sacrificed for the greater good starts to resonate with unacceptable totalitarian ideas, leading Tom Regan (1983) to dismiss the land ethic's disregard of the rights of the individual as "environmental fascism." More sympathetic readers of Leopold, however, have argued that a workable version of the land ethics maintains simply that both the biotic community and its individual members all have intrinsic value (Callicott 1989, 1999).

The idea that nature, its systems, populations, and communities have value in their own right is now so widely accepted that it is easy to forget just how radical a thought this was merely 20 or 30 years ago. And while the notion that economic value is fundamental or has priority over other values may seem to be widely rejected nowadays, there is still a widespread form of sickness and malaise that threatens the best efforts of environmentalists, conservationists, and others concerned about the increasing appropriation of the planet's biosynthetic product by human beings and the likelihood of catastrophic climate change induced by industrial societies. The term "affluenza" (James 2007) usefully captures the involuntary nature of this new malaise, conceived as a kind of infection generated by our exposure to the toxic structures of contemporary capitalism and consumerism.

Affluenza is a product of systems that are rich in available resources, hence able to thrive and multiply on the addictions cultivated in consumers. At the same time, these addictions are focused by the market on a range of pleasures and satisfactions that are—in comparison to the capacities of normal human beings—relatively narrow. The result is an atrophying of higher human sensitivities, a loss of the richness of lived experience, and its replacement by repetitive and short-lived satisfactions based on material goods. Alongside this are symptoms of addiction—typically a failure to recognize the scale of the problems we face, inflated optimism about our ability to get out of deeper and deeper problems, borrowing from the future, and an unrealistic and exaggerated sense of future profits and satisfactions. The problem of affluenza is not one that can be solved by reason or rational argument. Instead, it challenges policy makers to find ways of dealing with irrational and self-destructive addictions and to find ways of motivating new behaviors.

It is generally true that the longer it takes for people to master something, the more difficult it is for them to unlearn it. Likewise, the more time and effort that one has to invest in acquiring the taste for a certain kind of pleasure, the longer it will take for one to get bored with the experience. That explains why people become dissatisfied with consumer goods so quickly. Consumer goods, as such, are always updateable, upgradeable, upscalable. It is in the fate of any consumer product to become unsatisfying and therefore need replacement. They are not meant to last or provide long-lasting satisfaction. Ideal consumer products are those that people initially lust after, and then quickly become dissatisfied or bored with, and therefore soon seek a better replacement. That is how they keep the system of consumption and production running. That is why they are not intrinsically satisfying. Like junk food, consumer goods are immediately attractive, easy to enjoy, dangerously addictive, but quickly become dull, and harmful when consumed excessively. One way to counter affluenza, therefore, would be to cultivate in people the abilities to derive long-lasting pleasure and satisfaction from things and experiences that are simple and inexpensive, but the acquisition of the taste for which requires serious time and effort.

Leopold's seminal work may itself contain similar seeds of a response to the challenges of affluenza and consumer addiction. In a short piece on the sky dance of woodcock, he comments on the endless fascination and marvelling that we can enjoy when we are engrossed in and appreciative of our natural surroundings. "The drama of the sky dance," he writes, "is enacted nightly on hundreds of farms, the owners of which sigh for entertainment but harbor the illusion that it is to be sought in theaters. They live on the land, but not by the land" (Leopold 1949, p. 134). Here Leopold is contrasting the market-place pleasures of the entertainment industry (he is writing before television) with the endless joy that emerges from attentiveness to our surroundings and to the other beings with whom we share them. There may be resources here on which philosophers, analysts, and policy makers can draw when exploring ways in which to meet the challenges of affluenza, which undermines the capacity of the planet to maintain critical natural capital and the very systems on which human life and consumption themselves depend.

Leopold's genius in this case is not that he states a new idea, but that he gives concrete expression to ideas that had been mooted by John Muir early in the century and even earlier by writers such as John Stuart Mill. In his classic essay on *Utilitarianism* (1861), Mill wrote that "men lose their high aspirations as they lose their intellectual tastes, because they have not time or opportunity for indulging them." With striking insight into the consumer society still to emerge, Mill continued by commenting that "they addict themselves to inferior pleasures, not because they deliberately prefer them, but because they are either the only ones to which they have access, or the only ones which they are any longer capable of enjoying." Leopold's sketch of the sky dance of the woodcock and his comments on the puzzles, joy, and fascination it evokes remind the reader that humans still have the capacity to enjoy intrinsically worthwhile and valuable experiences provided they give themselves time to cultivate their own humanity within the broader environmental home where all life is set.

References

Allen, T.F.H., and T.W. Hoekstra. 1992. Toward a Unified Ecology. New York, NY: Columbia University Press.

Angermeir, P.L., and J.R. Karr. 1994. Biological integrity versus biological diversity as policy directives: Protecting biotic resources. Bioscience 44:690-697.

Callicott, J.B. 1989. In Defense of the Land Ethic: Essays in Environmental Philosophy. Albany, NY: State University of New York Press.

Callicott, J.B. 1995. The value of ecosystem health. Environmental Values 4:345-361.

Callicott, J.B. 1999. Beyond the Land Ethic: More Essays in Environmental Philosophy. Albany, NY: State University of New York Press.

James, Oliver. 2007. Affluenza. London: Vermilion.

Leopold, Aldo. 1939. The farmer as a conservationist. American Forests 45:294-99,316,323.

Leopold, Aldo. 1941. Wilderness as a land laboratory. Living Wilderness 6(1941).

Leopold, Aldo. 1949. A Sand County Almanac. Oxford: Oxford University Press.

Macdonald, David W., and Katrina Service. 2007. Key Topics in Conservation Biology. Oxford: Blackwell.

Markert, Bernd, and Jörg Oehlman. 1998. Ecotoxicology. *In* Modern Trends in Ecology and Environment, ed. R.S. Ambasht, 47-52. Leiden, the Netherlands: Backhuys Publishers.

Mill, John Stuart. 1861. Utilitarianism. *In* Collected Works of John Stuart Mill, Vol. 10, ed. J.M. Robson. Toronto: University of Toronto Press.

Regan, T. 1983. The Case for Animal Rights. London: Routledge & Kegan Paul.

Rolston, Holmes. 1992. Ethical responsibilities toward wildlife. Journal of the American Veterinary Medical Association 200:618-622.

ABOUT THE AUTHOR

Y.S. Lo, Lecturer, Department of Philosophy, La Trobe University, Victoria, Australia.

Resource Ownership and Property Rights

Kristin Shrader-Frechette

The land is in trouble. Water, air, and other resources have been so contaminated by human activities that they often cannot support life. The US Environmental Protection Agency (EPA) says approximately 56% of estuarine bodies, 45% of lakes, and 35% of rivers and streams in the nation are impaired—"unfit" for drinking, swimming, or fish consumption. This impairment comes mostly from pollutants like mercury, PCBs, chlordane, and dioxins, from sources like coal-fired plants, incinerators, and industrial stacks (EPA 1998). Only two US states (Wyoming and Alaska) have no restrictions on eating local fish because of their waterborne contamination (EPA 2004).

Wherever the land is in trouble, all living beings also are in trouble. Pollution contamination of air, water, and other resources, particularly since the Industrial Revolution, has harmed not only the land but all creatures who live on it. Studies done by the US National Academy of Sciences and the World Health Organization (WHO)—and published in medical journals like *Lancet*—show that increased environmental pollution causes increased cancer, heart trouble, respiratory diseases, asthma, allergies, and other ailments (Institute of Medicine 2005; Davis and Webster 2002). Apart from health effects caused by other air, water, and food pollutants, *Lancet* authors say that particulate air pollution alone annually causes 6.4% of children's deaths, ages 0 to 4, in developed nations. In Europe, this means that air particulates alone kill 14,000 toddlers each year (Valent et al. 2004). The WHO estimates that about 3 million people die prematurely each year because of both indoor and outdoor air pollution—more than 8,000 deaths each day (WHO 2005; Commonwealth Scientific and Industrial Research Organization 1999). One European Union scientific study calculated that between 25% and 40% of all UK deaths were "thought to be directly attributable to the effects of air pollution" alone (Gosline 2004), and similar conclusions hold for most of the industrialized world. After all, a recent *New England Journal of Medicine* long-term study of 90,000 twins (aimed at distinguishing genetically-based from environmentally-induced cancers) drew a clear conclusion. Although cigarette smoke is an environmental contaminant, and it causes about 30% of cancers, nevertheless other environmental pollutants cause most cancer problems; "the overwhelming contribution to the causation of cancer in the population of twins that we studied was the environment" (Lichtenstein 2002). Decades earlier, the US Office of Technology Assessment gave a similar warning; up to 90% of all cancers are "environmentally induced and theoretically preventable" (Lashof 1981). Because about 600,000 people in the United States die of cancer each year, this means that up to 540,000 of those deaths are environmentally induced and theoretically preventable (Shrader-Frechette 2007).

Flawed Property Rights Cause Land Contamination

One of the main reasons that land has been so massively misused and contaminated is that it has often been interpreted in terms of private, rather than the public, interest. Part of the reason for this individualistic emphasis may be that people have been insensitive to the ways in which land ownership confers political and economic power that can threaten both environmental welfare, justice, and the public good. Every public good, however, is bought at a price. And part of the price of protecting land is land-use controls and greater restrictions on property rights in natural resources. At least in the United States, however, property rights are often held more sacrosanct than civil rights.

This article makes two arguments: (1) Procedural justice—fairness in the procedures according to which people interact and exchange goods (see Rawls 1971)—requires that we restrict property rights in natural resources. (2) Conditions imposed by Locke's political theory—by expanding population, and by duties to future generations—require, in general, that we restrict property rights in natural resources such as land. If these two arguments are correct, we must restructure land ownership and use in a far more radical way than has ever been undertaken in the past.

The First Argument and the Plight of Farmers

To see why procedural justice requires restricting property rights in natural resources, consider the small farmer in California and in Appalachia. California agricultural land presents an important illustration of the need for land-use controls because owning even a small piece of it may confer a great deal of economic and political power. California is the largest producer of many specialized crops, and ownership of several hundred acres of farmland with rare soil and a specific climate can give one a great amount of power, for instance, in setting the price of broccoli or asparagus (Oxfam 2006; US Department of Commerce 1989, pp. 366, 356, 341-343; Caldwell and Shrader-Frechette 1993, pp. 65-84). A 1987 government study revealed that 67% of California farmland is held in farms that are larger than 2,000 acres each, and this concentration has only gotten worse in the last two decades. Another report on land ownership in California revealed, moreover, that 45 corporate farms, representing less than one-tenth of 1% of the commercial farms in the state, control approximately 61% of all farmland in California (Oxfam 2007; Shrader-Frechette 2002; Womack and Traub 1987, p. 12; Fellmeth 1973).

What is particularly disturbing is that highly concentrated ownership of California land has resulted in the owners having more concentrated political-economic power and greater ability to oppose contrary interests. As a consequence, large and small land owners do not have equal opportunity in the marketplace. And if they do not have equal opportunity, the demands of justice may not be met in situations where they compete in the same agricultural and economic markets (Binswanger 2007; Palmberg 2007). One practical reason that small farmers cannot compete equally has to do with inflated land values in California. Such land values benefit the large holders who drive up land prices which, in turn, hurt the small farmers who attempt to do all or most of their own work (Glasmeier and Farrigan 2003; see also Friedberger 1988; Galeski and Wilkening 1987; Davidoff and Gold 1979). If they are to compete with the larger holders, small farmers must continually have more land. But inflated land values make having more land more difficult, and expansion pressure inflates real-estate values further.

Since 1982, the number of California farms under 180 acres increased by less than 1%, whereas the number of California farms larger than 180 acres grew by 4.8%; family farms in California decreased by 0.8%, while the number of corporate farms increased by 10.7%; the number of farms operated by blacks in California has decreased by 13%, and the number operated by native Americans has decreased by 4%—mainly because small and minority farmers face more credit problems and initially have less land (US Department of Commerce 1989, pp. 1, 380; see also Oxfam 2007; Carlin and Mazie 1990). Because of both intentional and unintentional discrimination, small farmers, especially poor and minority farmers, are unable to accumulate the resources of land and credit that would give them opportunities equal to those of corporate farmers.

Something similar happens in Appalachia—where land has increased in value primarily because of its vast coal reserves. Here, most rural land is concentrated in the hands of a few corporate holders, a concentration that, as later paragraphs argue, has caused unequal political and economic opportunity. A major study of land ownership patterns and their impacts on the small farmer was sponsored several decades ago by the Appalachian Regional Commission and is one of the most comprehensive land-ownership studies ever completed in the United States. Presenting profiles of 80 counties in Alabama, Kentucky, North Carolina, Tennessee, Virginia, and West Virginia, the study concluded that most Appalachian woes—poverty (Ziliak 2007; Wimberley 2003), the decline of small farmers, housing shortages, and environmental degradation—were caused by concentrated, absentee ownership of most land. The researchers discovered that most owners of mineral rights paid less than a dollar an acre in property taxes; three-fourths paid less than 25 cents. They showed that 53% of total land surface in the 80 counties was controlled by 1% of the total population, and three-fourths of the surface acres surveyed were absentee-owned (outside the state); four-fifths were owned by out-of-state/county owners. Of the top 50 private owners, 46 were corporations (Egerton 1981, p. 43; see also Strange 1988; Whyte 1987; Gaventa and Horton 1981)—a situation that remains largely unchanged today (Billings 2004; Glasmeier and Farrigan 2003).

Using more than 100 socioeconomic indicators, the Appalachian land-use researchers concluded that (1) the greater the concentrated ownership of land/mineral resources, and the greater the absentee ownership, the less coal money remains in coal counties; (2) little land is owned by, or accessible to, local people; and (3) because of conclusions 1 and 2, many ills plague Appalachia: inadequate tax revenues/services; little economic development/diversified job opportunities; losses of agricultural lands; insufficient housing; little locally controlled capital; and outmigration rates from Appalachia proportional to corporate ownership and to land-ownership concentration (Gaventa and Horton 1981, pp. 210-212). In Kentucky, the study showed 58% of farms were under 99 acres; 62% in Tennessee (Womack and Traub 1987, pp. 36, 86). The researchers showed that concentrated absentee ownership of coal land caused virtually all Appalachian social-economic ills. Consequently many researchers say land reform/land-use controls are necessary, though not sufficient, conditions to correct these ills (Hollis 2006; Billings 2004; Glasmeier and Farrigan 2003; Gaventa and Horton 1981, p. 212; see also Mike Clark, Director of the Highlander Center, New Market, Tennessee, quoted by Egerton 1981, p. 44; Moss 1977; Griffin 1976; Albrecht and Murdoc 1990; Caldwell and Shrader-Frechette 1993, pp. 65-84).

Are there important ethical grounds for limiting property rights in natural resources so as to allow land-use controls? The basic premises of such an argument follow:

1. Concentrated ownership of Appalachian coal land and California agricultural land leads to concentrated political, legal, and economic power in the hands of the owners.
2. Such concentrations of political, legal, and economic power limit the voluntariness of land transactions (as well as other market transactions) between the large owners (holders of power) and others, especially small farmers.
3. Apart from legitimate reparation or punishment, whatever social institutions limit the voluntariness of market transactions (between large property owners and others) limit procedural justice.
4. Whatever limits procedural justice should be avoided.
5. Concentrated ownership of land ought to be avoided.
6. The only way to avoid concentrated ownership of land is to restrict property rights in land (Caldwell and Shrader-Frechette 1993, pp. 65-84).

Premises 1 and 2 above are largely factual claims whose truth has been documented in the land-ownership studies already cited. There also are intuitive reasons for believing that premises 1 and 2 are correct. One reason is that monopolies reduce the freedom of market transactions. The other insight is that property generates inequality, and inequality menaces liberty. Land economists, in particular, have explicitly noted how concentrations of rural land in the hands of a few owners leads to monopsony (owners' control of wages), the absence of developable land, a diversified economy, and local capital (see Deininger and Binswanger 1999; Griffin 1979; Griffin 1976). These factors in turn limit the voluntariness of transactions between large land owners and others (Pennock 1980; see Deininger and Binswanger 1999). Often small land owners are forced because of unfair competition to sell their lands/resources, at a disadvantage, to concentrated corporate farmers or coal companies—yet they cannot purchase land for farming. Because such unfair transactions limit voluntariness, they limit fairness and a free, competitive market necessary for procedurally just transactions. For Robert Nozick, John Rawls, and virtually all moral thinkers, procedural justice requires fairness; yet fairness requires conditions like voluntary transactions among people (see Nozick 1974, pp. 90-93; Rawls 1971).

Using the case of the small California or Appalachian farmer, premises 1 and 2 of the argument so far have suggested that factors like monopsony and limited local capital coerce small farmers, so that their land transactions likely are nonvoluntary. Of course, the obvious objection is that such a claim makes the conditions of morality (e.g., voluntariness) unrealistic, because many choices may be nonvoluntary. However, this objection fails because most contemporary choices are not made in contexts of serious threats to well-being, yet small California and Appalachian farmers face serious threats; indeed, Appalachia is one of the poorest US regions. Also, many alternatives (to people's choices) are not all undesirable, yet severe poverty, especially in Appalachia (Ziliak 2007; Wimberley 2003), makes many land-related choices undesirable and makes their unequal opportunity even worse.

Regarding premise 3, concentrated Appalachian property holdings limit the voluntariness and hence fairness and procedural justice of many land transactions. Concentrated holdings cause choices of those "less propertied" often to be made under threat, compulsion, and without good alternatives. In Appalachia, where most of the land is held by few individuals, their speculation has driven up land prices and impeded settlement by poor Americans. Because of limited Appalachian industry and rural population, small farmers have rarely been able to compete fairly with large landowners—who likely own the community bank and general store and thus control the community. Chronically impoverished, poorly educated, and victimized by flawed tax laws, small Appalachian farmers face great difficulties when competing for land with large farmers and coal companies. They have little capital, needed to keep their land, and there are few other (i.e., nonagricultural, nonmining) local jobs. Thus, because they have unequal bargaining power and opportunity, they likely sell neither their land nor workplace skills voluntarily (Held 1976; see also Gragson and Bolstad 2006; Caldwell and Shrader-Frechette 1993, pp. 65-84; Sterba 1980).

Regarding premise 4 and subsequent premises, their truth is intuitively obvious, once the earlier factual premises are granted. Note, however, that the preceding or first argument for land-use controls does not rely on any socialistic justifications, because it specifies no particular distribution of land as desirable. Instead it presupposes only that land-use controls (and property-rights restrictions) must be proportionate to the degree of procedural injustice they prevent, and to the degree of equal opportunity they promote.

A Second Argument for Limiting Property Rights in Natural Resources

Apart from preceding considerations of procedural justice, a second argument for limiting property rights focuses on human labor. This argument is that traditionally only labor (or what is exchanged for labor) gives property rights, yet because natural resources are not created by human labor or by what is exchanged for labor (like money), there are no full property rights to them (Caldwell and Shrader-Frechette 1993, pp. 65-84). Rather, one has property rights only to the value, added by labor, to natural resources (Locke 1984; Nozick 1984; Locke 1980; Becker 1977; Nozick 1974, pp. 174-178; Caldwell and Shrader-Frechette 1993, pp. 65-84). Thus, alleged natural-resource owners have no full property rights to them.

A related reason to doubt full property rights to land is that implementing these rights often involves exhausting a common resource (e.g., coal) by a subset of the population. Yet Locke's theory of property rights is acknowledged in the West to be the foundation of property rights, and Locke stipulates that one may own/appropriate property only if "as much and as good" is left for others. One may not take or retain commons/resource property, if doing so threatens others' equal opportunity to use/enjoy the resource (Caldwell and Shrader-Frechette 1993, pp. 65-74; see also Locke 1984; Locke 1980; Becker 1977).

The second (or labor) argument is open to several objections. One is that (1) although one's labor does not create value in natural resources, utilitarian arguments support property rights in natural resources; and another is that (2) there is no reason why the industrious should not gain competitive advantages over the nonindustrious. Consider each of these objections. Robert Nozick formulates objection 1, admitting grounds for denying property rights in natural objects, but arguing that "social considerations" favor private property in resources. These social considerations include putting the means of production in the hands of those who can use them most efficiently; encouraging experimentation; and leading some people to hold back resources, from current consumption, for future markets (Nozick 1984, p. 148). Nozick's objection is that although one cannot rationally justify property rights in natural resources, one can do so on utilitarian grounds (see Bardon 2002).

The Nozick objection 1 errs partly because he suggests property rights in resources encourage experimentation. Yet if people are eager to use resources profitably, this desire can contradict tendencies to experiment. Also, if land is a common resource, why should any single person have the right to experiment with it? And how could private property protect future generations, by leading some persons to hold back resources from current consumption? The dire water and air-pollution statistics, given earlier, suggest little "holding back." They suggest exponential resource use, rather than leaving profits for others (Meadows 1992; Brown 2002; see Shrader-Frechette 2007, 1991; Meadows 1974). Nozick likewise errs in presupposing property rights put resources in the hands of those who can use them most efficiently. This presupposition erroneously assumes that natural resources can be employed for private gain, although private labor did not create them. It assumes economic efficiency outweighs equality and rights of future generations. Nozick thus begs the very question at issue: whether there ought to be property rights in natural resources. It makes no sense to say a private individual ought to be allowed to maximize the economic efficiency of something—unless he antecedently has property rights over the "something"—and yet Nozick gives no reason for believing he has those rights.

Nozick's objection 2 also errs because he assumes property rights in resources gives advantages to the industrious over the nonindustrious. However, if an industrious person obtains property in natural resources because of his hard work, intelligence, and ambition, it is not clear that he has won something "away from" the lazy, unintelligent, and unambitious. He may have won something from future generations. Even if one concedes (erroneously) that property rights in natural resources allows the industrious to be rewarded over the nonindustrious, why should the aggressive inherit the earth? Why should the natural advantage of intelligence allow one to receive greater benefits from the commons? Besides, to the extent that property rights protect possession and inheritance, the strong may not need them. The weak do (see Becker 1977, p. 44; Rashdall 1915). If so, there are no clear grounds for arguing that already-strong persons have rights to property in natural resources.

Conclusion

The preceding arguments urge limiting property rights in natural resources, so as to avoid unfair competition, especially in resource-rich areas like Appalachia and California, and to avoid claiming property rights to things not created by one's labor. If these ethical arguments are right, severely limiting property rights in natural resources might provide a foundation for avoiding the land destruction outlined at the beginning of this paper.

References

Albrecht, D.E., and S.H. Murdoc. 1990. The Sociology of US Agriculture. Ames, IA: Iowa State University Press.

Bardon, Adrian. 2002. From Nozick to welfare rights: Self ownership, property, and moral desert. Critical Review 14(4):481-501.

Becker, L. 1977. Property Rights. London: Routledge and Kegan Paul: 43-5.

Billings, Dwight B. 2004. Social origins of Appalachian poverty: Markets, cultural strategies, and the state in an Appalachian Kentucky community, 1804-1940. Rethinking Marxism 16(1):19-36.

Binswanger, Hans P. 2007. Empowering rural people for their own development. Agricultural Economics 37(1):13-27.

Brown, Charles E. 2002. World Energy Resources. Berlin, Germany: Springer-Verlag: 20-21.

Caldwell, Lynton, and Kristin Shrader-Frechette. 1993. Policy for Land. Lanham, MD: Rowman and Littlefield.

Carlin, T.A., and S.M. Mazie. 1990. The US Farming Sector Entering the 1990's. Washington, DC: US Department of Agriculture.

Commonwealth Scientific and Industrial Research Organization. 1999. Air Pollution. Aspendale, Australia: Commonwealth Scientific and Industrial Research Organization. http://www.csiro.au/index.asp?id=AirPollution&type=mediaRelease.

Davidoff, Paul, and N. Gold. 1979. The supply and availability of land for housing for low- and moderate-income families. In Land-Use Controls, ed. David Listokin. New Brunswick, NJ: Rutgers University Press: 279-284.

Davis, D.L., and P.S. Webster. 2002. The social context of science: Cancer and the environment. Annals of the American Academy of Political and Social Science 584(Nov 2002):13-34.

Deininger, Klaus, and Hans Binswanger. 1999. The evolution of the World Bank's land policy: Principles, experience, and future challenges. The World Bank Research Observer 14(2):247-276.

Egerton, John. 1981. Appalachia's absentee landlords. The Progressive 45(6):43-44.

EPA (Environmental Protection Agency). 1998. National Water Quality Inventory. Washington, DC: EPA: ES-3. http://www.epa.gov/305b/98report/.

EPA. 2004. National Listing of Fish Advisories. Washington, DC: EPA: 2.

Fellmeth, R.C. 1973. The Politics of Land. New York: Grossman: 12.

Friedberger, M. 1988. Farm Families and Change. Lexington, KY: University Press of Kentucky: 73, 223ff.

Galeski , B., and E. Wilkening, eds. 1987. Family Farming in Europe and America. Boulder, CO: Westview.

Gaventa, John, and Bill Horton. 1981. Ownership Patterns and Their Impacts on Appalachian Communities: A Survey of 80 Counties, vol. 1. Washington, DC: Appalachian Regional Commission: 25-29, 210-212.

Glasmeier, Amy K., and Tracey L. Farrigan. 2003. Poverty, sustainability, and the culture of despair: Can sustainable development strategies support poverty alleviation in America's most environmentally challenged communities? Annals of the American Academy of Political and Social Science 590(1):131-149.

Gosline, Anna. 2004. European deaths from air pollution set to rise. New Scientist 17(51)(Sept 2004). http://www.newscientist.com/article.ns?id=dn6364).

Gragson, Ted. I., and Paul V. Bolstad. 2006. Land use Legacies and the future of southern Appalachia. Society & Natural Resources 19(2):175-190.

Griffin, K. 1976. Land Concentration and Rural Poverty. New York: Holmes and Meier: 1-11.

Griffin, K. 1979. The Political Economy of Agrarian Change. London: Macmillan: 40, 223-225.

Held, Virginia. 1976. John Locke on Robert Nozick. Social Research 43(1):171-172.

Hollis, Paul L. 2006. Alabama growers look to protect property rights. Southeast Farm Press 33(4):13-16.

Institute of Medicine and Lovell Jones, John Paretto, and Christine Coussens, eds. 2005. Rebuilding the Unity of Health and the Environment. Washington, DC: National Academy Press: 2, 15, 43-44.

Lashof, J.C., et al., Health and Life Sciences Division of the US Office of Technology Assessment. 1981. Assessment of Technologies for Determining Cancer Risks from the Environment. Washington, DC: Office of Technology Assessment: 3, 6ff.

Lichtenstein, Paul, Niels Holm, Pia Verkasalo,Anastasia Iliadou, Jaakko Kaprio, Markku Koskenvuo, Eero Pukkala, Axel Skytthee, and Kari Hemminki. 2002. Environmental and heritable factors in the causation of cancer. New England Journal of Medicine 343(2): 78-85.

Locke, John. 1980. Second Treatise of Government. Hackett Publishing: ch. 5, par. 27.

Locke, John. 1984. Of property. In Property, ed. L. Becker and K. Kipnis. Englewood Cliffs, NJ: Prentice-Hall: 138.

Meadows, D.H., D.L. Meadows, and J. Randers. 1992. Beyond the Limits. Post Mills, VT: Chelsea Green Publishing Company: chs. 2, 4, 6.

Meadows, D.H., et al. 1974. The Limits to Growth. New York: New American Library: 40, 60, 69, 81.

Moss, Blaine, Natural Resources Defense Center, 1977. Land Use Controls in the United States. New York: Dial Press/James Wade: 235-236.

Nozick, Robert. 1974. Anarchy, State, and Utopia. New York: Basic Books: 90-93, 174-178.

Nozick, Robert. 1984. Locke's theory of acquisition. In Property, ed. L. Becker and K. Kipnis. Englewood Cliffs, NJ: Prentice-Hall: 146-149.

Palmberg, Elizabeth. 2007. Planting justice for farmers. Sojourners Magazine 36(7):8.

Pennock, J.R. 1980. Thoughts on the right to private property. In Property, Nomos 22, ed. J.R. Pennock and J.W. Chapman. New York: New York University Press: 269.

Rashdall, Hastings. 1915. The philosophical theory of property. In Property: Its Duties and Rights, 2nd ed., ed. J.V. Bartlett. London: Macmillan: 54-56.

Rawls, John. 1971. A Theory of Justice. Cambridge, MA: Harvard University Press: 111-113, 342-347.

Shrader-Frechette, K. 1991. Environmental Ethics. Pacific Grove, CA: Boxwood: 171ff.

Shrader-Frechette, K. 2002. Environmental Justice: Creating Equality, Reclaiming Democracy. New York: Oxford University Press: 51-53.

Shrader-Frechette, K. 2007. Taking Action, Saving Lives. New York: Oxford University Press.

Sterba, James P. 1980. Neo libertarianism. *In* Justice: Alternative Political Perspectives, ed. J.P. Sterba. Belmont, CA: Wadsworth.

Strange, M. 1988. Family Farming. Lincoln, NE: University of Nebraska Press: 171, 199-200.

US Department of Commerce, Bureau of the Census. 1989. 1987 Census of Agriculture 1, no. 5, California. Washington, DC: US Government Printing Office: 1, 341-343, 356, 366, 380.

Valent, F., D.A. Little, R. Bertollini, L E. Nemer, G. Barbonc, and G. Tamburlini. 2004. Burden of disease attributable to selected environmental factors and injury among children and adolescents in Europe. Lancet 363:2032-2039.

WHO (World Health Organization). 2005. Indoor Air Pollution and Health. Bonn: WHO. http://www.who.int/mediacentre/factsheets/fs292/en/.

Whyte, W., ed. 1987. Our American Land. Washington, DC: US Government Printing Office: 122ff.

Wimberley, Ronald C. 2003. U.S. poverty in space and time: Its persistence in the South. Sociation Today 1(2)(Fall 2003). http://nsdl.org/resource/2200/20061003150842416T.

Womack, L.M., and L.G. Traub. 1987. US–State Agricultural Data. New York: US Department of Agriculture, AIB 512: 12, 36, 86.

Ziliak, James P. 2007. Human capital and the challenge of persistent poverty in Appalachia. Economic Commentary (Feb 2007):1-4.

ABOUT THE AUTHOR

Kristin Shrader-Frechette, O'Neill Family Endowed Professor, Department of Philosophy and Department of Biological Sciences, University of Notre Dame, Notre Dame, Indiana.

Living with the Land

Wendell Berry

Editor's note. This piece was presented by Wendell Berry as the keynote address at the 1991 Soil and Water Conservation Society annual conference in Lexington, Kentucky.

Though we still live from the land, as we always have and always must, we now live with the land less than ever before. To live with something is, by implication, to live intimately or companionably with it. This relation implies mutuality of interest, interdependence, familiarity, affection, care. When herbicides fall with the rain, we are living with chemistry, not with the land.

In our relation to the land, we are ruled by terms and limits set, not by anyone's preference, but by nature and by human nature:

- Land that is used will be ruined unless it is properly cared for.
- Land cannot be properly cared for by people who do not know it intimately, who do not know how to care for it, who are not strongly motivated to care for it, and who cannot afford to care for it.
- People cannot be adequately motivated to care for land by general principles or by incentives that are merely economic. That is, they won't care for it merely because they think they should do so or merely because somebody pays them to do so.
- They are motivated to care for land—to live with it—insofar as their interest in it is direct, dependable, and permanent.
- They will be motivated to live with the land if they can reasonably expect to live on it as long as they live. They will be more strongly motivated if they can reasonably expect that their children and grandchildren will live on it as long as they live. That is, there must be a mutuality of belonging. They must feel that the land belongs to them, that they belong to it, and that this belonging is a settled and unthreatened fact.
- That this belonging must be appropriately limited is the indispensable qualification of the idea of land ownership. It is well understood that ownership is an incentive to care. But there is a limit beyond which an owner cannot take proper care of land that he or she is using. The need for attention increases with the intensity of use. But the quality of attention decreases as acreage increases.
- A nation will destroy its land and, therefore, itself if it does not foster in every possible way the sort of thrifty, prosperous, permanent rural households and

142

communities that have the desire, the skills, and the means to care properly for the land they are using.

In an age notoriously impatient of restraints, such a list of rules will hardly be welcome, but that these are the rules of land use I have no doubt. I am convinced of their authenticity both by common wisdom and by my own experience and observation. The rules exist; the penalties for breaking them are obvious and severe; the failure of land stewardship in this country is the result of a general disregard of them all.

No Reason for Optimism

In proof of this failure, there is no use in reciting again the statistics of land ruination. The gullies and other damages are there to be seen. Little of our land that is being used—for logging, mining, or farming—is being well used. Much of our land has never been well used. Those of us who know what we are looking at know this is true. And we know there is little reason for optimism.

After observing the everywhere worsening condition of our land, we only have to raise our eyes a little to see the worsening condition of those who are using the land and are entrusted with its care. We must accept as fact that, by now, our country (as opposed to our nation) is characteristically in decline. War, depression, inflation, usury, the attitudes of the industrial economy, social and educational fashions—all have taken their toll.

For a long time the news from everywhere in rural America has been almost unrelievedly bad: bankruptcy, foreclosure, depression, suicide, the departure of the young, the loneliness of the old, soil loss, soil degradation, chemical pollution, loss of genetic and specific diversity, extinction or threatened extinction of species, depletion of aquifers, stream degradation, loss of wilderness, strip mining, clearcutting, population loss, loss of supporting economies, the deaths of towns. Rural American communities, economies, and ways of life that in 1945 were thriving and, though imperfect, full of promise for an authentic human settlement of our land, are now as effectively destroyed as the Jewish communities of Poland; the means of destruction were not as direct, but they have proved just as thorough.

The news of rural decline and devastation has been accompanied by a chorus of professional, institutional, and governmental optimists, insisting that all was well; that farmers who failed were merely "inefficient producers" for whose failure the country was better off; that money and technology would fill the gaps; that government would fill the gaps; that science would soon free us from our regrettable dependence on the soil; that science, in fact, would soon perfect an end run around both nature and human depravity. We have heard that it is good business and good labor economics to destroy the last remnants of American wilderness. We have heard that the rural population is actually growing because city people are moving to the country and commuters are replacing farmers. We have heard that the rural economy can be repaired by moving the urban economy out into the country and thus replacing rural work with work in factories and offices. Meanwhile, the real conditions of rural land and rural people have been getting worse.

One Example

Of the general condition of the American countryside, my own community will serve well enough as an example. Port Royal, Kentucky, has a population of perhaps 125. The town came into existence as a trading center, serving the farms in a few square miles of hilly countryside on the west side of the Kentucky River. It has never been much bigger than it is now. But whereas now it is held together by habit or convenience, it once was held together by a complex local economy. In my mother's childhood, in the years before World War I, there were 16 business and professional enterprises in the town, all serving the town and surrounding farms. By the time of my own childhood, in the years before World War II, that number had been reduced to 12, but the town and its tributary landscape were still alive as a community and as an economy. Now, counting the post office, the town has five enterprises, one of which does not serve the local community. There is now no market for farm produce in the town or within 40 miles. We no longer have a garage or repair shop of any kind. We have had no doctor for 40 years, and no school for 30. As a local economy and, therefore, as a community, Port Royal is dying.

What does the death of a community, a local economy, cost its members? What does it cost the country? So far as I know, we have no economists who are interested in such costs. Nevertheless, when you must drive 10, 20, or more miles to reach a doctor or a school or a mechanic, or to find parts for farm machinery, the costs exist, and they are increasing. As they increase, they make the economy of every farm and household less tenable.

As people leave the community or, remaining in the place, drop out of the local economy, as the urban-industrial economy more and more usurps the local economy, as local economic constraints increase with the scale and speed of work, care declines. As care declines, the natural supports of the human economy and community also decline; whatever is used is used destructively.

We in Port Royal are part of an agricultural region surrounded by cities that import much of their food from distant places. Though we urgently need crops that can be substituted for tobacco, we produce virtually no vegetables or other foods for consumption in our region. Having no local food economy, we produce a less and less diverse food supply for the general market. This condition implies and virtually requires the abuse of our land and our people, and they are abused.

We are also part of a region that is abundantly and diversely forested, and we have no forest economy. We have no local wood products industry. This makes it almost certain that our woodlands and their owners will be abused, and they are abused.

We provide a great deal of recreation for our urban neighbors—hunting, fishing, boating, and the like—and we have the capacity to provide more. But for this we receive either almost nothing or nothing, and sometimes we suffer damage.

In our region, moreover, there has been no public effort to preserve the least scrap of land in its pristine condition. And the last decade or so of agricultural depression has caused much logging of the stands of mature forest in private hands. Now, if we want our descendants to know what the original forest was like—that is, to know the original nature of our land—we must start from scratch and grow the necessary examples over this next 200 or 300 years.

My part of rural America is, in short, a colony, like every other part of rural America. Virtually the whole landscape of our country—from the exhausted cottonfields of the plantation South to the eroding wheatlands of the Palouse, from the strip mines of Appalachia to the clearcuts on the Pacific slopes—is in the power of an absentee economy, once national and now increasingly international, that is without limit in its greed and without mercy in its exploitation of land and people. Between the prosperity of this vast central economy and the prosperity of any local economy or locality there is now a radical disconnection. The accounting that measures the wealth of corporations, great banks, and national treasuries takes no measure of the civic or economic or natural health of Port Royal, Kentucky, of Harpster, Ohio, or Indianola, Iowa, or Mattfield Green, Kansas, or Wolf Hole, Arizona, or Nevada City, California, and it does not intend to do so.

The Jeffersonian Dream

In 1912, according to William Allen White, "the county in the United States with the largest assessed valuation was Marion County, Kansas.… Marion County happened to have a larger per capita of bank deposits than any other American county.… Yet no man in Marion County was rated as a millionaire, but the jails and poorhouses were practically empty. The great per capita of wealth was actually distributed among the people who earned it."

This, of course, is the realization of that dream that is sometimes called Jeffersonian, but is really the dream of the economically oppressed throughout human history. Because this was a rural county, White was not talking just about bank accounts; he was talking about real capital, usable property. That era and that dream are now long past. Now, the national economy, which is increasingly a global economy, no longer prospers by the prosperity of the land and people, but by their exploitation.

The Civil War made America safe for the moguls of the railroads and the mineral and timber industries who wanted to be free to exploit the countryside. The work of these industries and their successors is now almost complete. They have dispossessed, disinherited, and moved into the urban economy virtually the entire citizenry; they have defaced and plundered the countryside. Now, this great corporate enterprise, thoroughly uprooted and internationalized, is moving toward the exploitation of the whole world under the shibboleths of "globalization," "free trade," and "new world order." The proposed revisions in the General Agreement on Tariffs and Trade are intended solely to further this exploitation. The aim is simply and unabashedly to bring every scrap of productive land and every worker on the planet under corporate control.

The voices of the countryside, the voices appealing for respect for the land and for rural community, have simply not been heard in the centers of wealth, power, and knowledge. The centers have decreed that the voice of the countryside shall be that of Snuffy Smith or Li'l Abner, and only that voice they have been willing to hear.

"The business of America is business," a prophet of our era too correctly said. The corollaries are clearly implied: that the business of the American government is to serve, protect, and defend business; that the business of the American people is to serve the government, which is to serve business. The costs of this state of things are incalculable. To start with, people in great numbers, because of their perception that the government serves not the country or the people but the corporate economy, do not vote—and thus vote against the government. Our leaders, therefore, are now in the curious and hardly legitimate position of asking a substantial number of people to cheer for, pay for, and perhaps die for, a government that they have not voted for.

But when the interests of local communities and economies are relentlessly subordinated to the interests of "business," then two further catastrophic results inevitably follow. First, the people are increasingly estranged from the native wealth, health, knowledge, and pleasure of their country. Second, the country itself is destroyed.

To Heal the Earth

It is not impossible to look at the present condition of our land and people and find support for optimism. We must not fool ourselves. It is altogether conceivable that we may go right along with this business of "business," with our curious religious faith in technological progress, with our glorification of our own greed and violence, always rationalized by our indignation at the greed and violence of others, until our land, ourselves, and our world are utterly destroyed. We know from history that massive human failure is possible. It is foolish to assume that we will save ourselves from any fate that we have made possible simply because we have the conceit to call ourselves *Homo sapiens*.

On the other hand, we want to be hopeful, and hope is one of our duties. A part of our obligation to our own being and to our descendants is to be always studying our life and our condition, searching for the authentic underpinnings of hope. And if we look, these underpinnings can still be found.

For one thing, though we have caused the earth to be seriously diseased, it is not yet without health. The earth we have before us now is still abounding and beautiful. We must learn again to see that present world for what it is. The health of nature is the primary ground of hope—if we can have the humility and wisdom to accept nature as our teacher. Nature is the best farmer and forester, for she does not destroy the land in order to make it productive. So in our wish to live with our land, we are not without the necessary lessons; nor are we without instruction, in our cultural and religious tradition, to learn those lessons.

But we have not just the example of nature; we have, still, though few and widely scattered, many examples of competent and loving human stewardship of the earth. We have too our own desire to be healthy in a healthy world. Surely, most of us still have, somewhere within us, the fundamental human wish to die in a world in which we have been glad to live. And we are, in spite of much evidence to the contrary, somewhat sapient. We can think—if we will. If we know carefully enough who, what, and where we are and if we keep the scale of our work small enough, we can think responsibly.

These assets are not the gigantic, technical, and costly equipment that we tend to think we need, but they are enough. They are, in fact, God's plenty. Because we have these assets, which are the supports of our legitimate hope, we can start from where we are, with what we have, and imagine and work for the healings that are necessary.

We must begin by giving up any idea that we can bring about these healings without fundamental changes in the way we think and live. We face a choice that is starkly simple: We must change or be changed. If we fail to change for the better, then we will be changed for the worse. We cannot blunder our way into health by the same sad and foolish hopes by which we have blundered into disease. We must see that the standardless aims of industrial communism and industrial capitalism have failed. The aims of productivity, profitability, efficiency, limitless growth, limitless wealth, limitless power, limitless mechanization and automation can enrich and empower the few (for awhile), but they will sooner or later ruin us all. The gross national product and the corporate bottom line are utterly meaningless as measures of the prosperity or health of the country.

If we want to succeed in our dearest aims and hopes as a people, we must understand that we cannot proceed any farther without standards, and we must see that ultimately the standards are not set by us but by nature. We must see that it is foolish, sinful, and suicidal to destroy the health of nature for the sake of an economy that is really not an economy at all, but merely a financial system, and that is unnatural, undemocratic, sacrilegious, and, on its own terms, ephemeral. We must see the error of our effort to live by fire, by burning the world in order to live in it. There is no plainer symptom of our insanity than our avowed intention to maintain by fire an unlimited economic growth. Fire destroys what nourishes it, and so, in fact, imposes severe limits upon any growth associated with it.

The true source and analogue of our economic life is the economy of plants, which never exceeds natural limits, never grows beyond the power of its place to support it, produces no waste, and enriches and preserves itself by death and decay. We must learn to grow like a tree, not like a fire. We must repudiate what Edward Abbey called "the ideology of the cancer cell": the idiotic ideology of "unlimited economic growth," which pushes blindly toward the limitation of massive catastrophe.

We must give up also our superstitious conviction that we can contrive technological solutions to all our problems. Soil loss, for example, is a problem that embarrasses all of our technological pretensions. If soil were all being lost in a huge slab somewhere, that would appeal to the would-be heroes of science and technology, who conceivably might engineer a glamorous, large, and speedy solution—however many new problems they might cause in doing so. But soil is not usually lost in slabs or heaps of magnificent tonnage. It is lost a little at a time over millions of acres by the acts or the carelessness of millions of people. It cannot be saved by heroic feats of gigantic technology, but only by millions of small acts and restraints, conditioned by small fidelities, skills, and desires. Soil loss is ultimately a cultural problem, and it will be corrected only by cultural solutions.

The aims of production, profit, efficiency, economic growth, and technological progress imply, as I have said, no social or ecological standards, and in practice they submit to none. But there is another set of aims that does imply a standard, and these aims are freedom (a synonym for personal and local self-sufficiency) and pleasure (our wish that human freedom and pleasure may last). The standard implied by all of these aims is health. They depend ultimately and inescapably upon the health of nature; the idea that freedom and pleasure can last long in a diseased world is preposterous. But these good things depend also upon the health of human culture, which is to a considerable extent the knowledge of economic and other domestic procedures—ways of work, pleasure, and education—that they preserve the health of nature.

The Idea of Community

In talking about health, we have thus begun to talk about community. But we must take care to see how this standard of health enlarges and clarifies the idea of community. If we speak of a healthy community, we cannot be speaking of a community that is merely human. We are talking about a neighborhood of humans in a place and the place itself: its soil, its water, its air, and all of the families and tribes of the nonhuman creatures that belong to it. If the place is well-preserved; if its entire membership, natural and human, is present in it; and if the human economy is in practical harmony with the nature of the place, then the community is healthy. A diseased community will be suffering natural losses that become, in turn, human losses. A good community is sustainable; it is within reasonable limits self-sufficient; and it is within reasonable limits self-determined (free of tyranny).

Community, then, is an indispensable term in any discussion of the connection between people and land. A healthy community would be a form that would include all of the local things that are connected by the larger, ultimately mysterious form of the Creation. In speaking of community, then, we are speaking of a complex connection, not only among human beings, or between humans and their homeland, but between the human economy and nature, between forest or prairie and field or orchard, and between troublesome creatures and pleasant ones. All neighbors are included.

From the standpoint of such a community, any form of land abuse—a clearcut, a strip mine, an overplowed or overgrazed field—is as alien and as threatening as it would be from the standpoint of an ecosystem, for such a community would be at one with the ecosystem. From such a stand standpoint, it would be plain that land abuse reduces the possibilities of local life, just as do chain stores, absentee owners, and consolidated schools.

One obvious advantage of such an idea of community is that it provides a common ground and a common goal between conservationists and small-scale land users. The long-standing division between conservationists and farmers, ranchers and other private, small business people is distressing because it is to a considerable extent false. It is readily apparent that the same economic forces that threaten the health of ecosystems and the survival of species are equally threatening to economic democracy and the survival of human neighborhoods.

Building Healthy Communities

I believe that the most necessary question, now—for conservationists; for small-scale farmers, ranchers, and business people; for politicians interested in the survival of democracy; and for consumers—is this: What must be the economy of a healthy community based in agriculture or forestry? It cannot be the present colonial economy in which only "raw materials" are exported and all necessities and pleasures are imported. To be healthy, land-based communities will need to add value to local products; they will need to supply local demand; and they will need to be reasonably self-sufficient in food, energy, pleasure, and other needs.

Once a person understands the necessity of healthy local communities and community economies, it becomes easy to imagine a range of reforms that might bring them into being.

It is at least conceivable that useful changes might be started or helped along by consumer demand in the cities. There is already evidence of a growing concern among urban consumers about the quality and purity of food. Once this demand grows extensive and competent enough, it will have the power to change agriculture—if there is enough left of agriculture, by then, to be changed.

It is even conceivable that our people in Washington might make decisions tending toward sustainability and self-sufficiency in local economies. The federal government could do much to help, if it would. Its mere acknowledgement that problems exist would be a promising start.

But let's admit that urban consumers are not going to be well-informed about their economic sources very soon and that a federal administration enlightened about the needs and problems of the countryside is not in immediate prospect.

Such an admission will discourage only those who think that no change can come that is not instigated in Washington, DC, and the other great urban centers of wealth and power. Others, I hope, will allow the possibility that change instigated in those places may be as undesirable as it is improbable. Those of us who are concerned for the fate of our land had better consider the likelihood that it cannot be conserved by more central planning by more absentee experts.

The real improvements must come to a considerable extent from the local communities themselves. We need local revision of our methods of land use and production. We need to cooperate and study and work together to reduce scale, overhead, and industrial dependencies; we need to market and process local products locally; we need to bring local economies into harmony with local ecosystems to substitute ourselves, our neighborhoods, our local resources for expensive imported goods and services; we need to increase cooperation among all local economic entities—households, farms, factories, banks, consumers, and suppliers. This is not just work for the privileged, the well-positioned, the wealthy, and the powerful. It is work for everybody. It needs the help of state, county, and city governments, of all kinds of working people, of people in neighborhoods and households, of individuals.

To advocate such reforms is to advocate a kind of secession—not a secession of armed violence, but a quiet secession by which people find the practical means and the strength of spirit to remove them from an economy that is exploiting them and destroying their homeland. The great, greedy, indifferent national and international economy is killing rural America—it is killing our country. Experience has shown that there is no use in appealing to this economy for mercy toward the earth or toward any human community. All true patriots must find ways of opposing it.

ORIGINAL PUBLICATION DETAILS

By Wendell Berry, Farmer, Writer, and Professor of English at the University of Kentucky, Port Royal, Kentucky.

Berry, Wendell. 1991. Living with the land. *Journal of Soil and Water Conservation* 46(6):390-393.